ML

AMONG OUR BOOKS

Publication in English

- Enhanced Oil Recovery
     M. Latil
     with the assistance of C. Bardon, J. Burger, P. Sourieau

- Properties of Reservoir Rocks: Core Analysis.
     R.P. Monicard

- Multiphase Flow in Porous Media.
     C.M. Marle

- Seabed Reconnaissance and Offshore Soil Mechanics for the Installation of Petroleum Structures.
     P. Le Tirant

- Drilling Data Handbook.

INSTITUT FRANÇAIS DU PÉTROLE
ÉCOLE NATIONALE SUPÉRIEURE DU PÉTROLE ET DES MOTEURS

Louis H. REISS
Senior Reservoir Engineer
with Elf Aquitaine and
Visiting Lecturer at
Ecole Nationale Supérieure du Pétrole et des Moteurs

THE RESERVOIR ENGINEERING ASPECTS OF FRACTURED FORMATIONS

Translation from the French
by Max CREUSOT
M.A. " Cantab "
Senior Reservoir Engineer
with Elf Aquitaine

1980

ÉDITIONS TECHNIP 27 RUE GINOUX 75737 PARIS CEDEX 15   **techni**

ISBN 2-7108-0374-7

Printed in France
by Imprimerie Louis-Jean, 05002 Gap

In memory of
Emmanuel Lefebvre du Prey

PREFATORY NOTE

Specialists from the petroleum industry, the *Institut Français du Pétrole* (*IFP*) the *Ecole Nationale Supérieure du Pétrole et des Moteurs* (*ENSPM*) and universities teach at the Graduate Study Center for Drilling and Reservoir Engineering of *ENSPM*.

In conjunction with this teaching, they have written various books dealing with the different scientific and technical aspects of these petroleum operations.

The present book is one of them.

foreword

This volume started as an in-house document within the *Société Nationale Elf Aquitaine—Production (SNEA-P)* to meet a gap in published textbooks on reservoir engineering. It was designed both as a training manual (and has been used in many of the courses given by the *Ecole Nationale Supérieure du Pétrole et des Moteurs (ENSPM)* since 1975) and as a reference for engineers working overseas. Its aim was to integrate both the practical experience and theoretical work available on fractured reservoirs. In this context the research carried out by the *Institut Français du Pétrole (IFP)* and the exchanges of ideas with Soviet experts under the auspices of the Franco-Soviet cooperation agreements have been invaluable. No book on fractured reservoirs could be complete without reference to the Iranian experience, and the author is in debt to the engineers with whom he has discussed these reservoirs.

Recently *IFP,* which has sponsored a number of textbooks on petroleum engineering, decided to print them in English. It was thought that this booklet would correspond to a need, and the author is grateful to :

Those who have authorized the publication of this document, the management of *SNEA-P.*

Those who have made possible this type of synthesis, the managers and research workers of the *Association de Recherche sur les Techniques d'Exploitation du Pétrole (ARTEP).*

Those who have helped and encouraged the author, and in particular D. BOSSIE-CODREANU, F. CRAIG, H. KAZEMI, C. MARLE, M. MOLLIER, A.M. SAIDI.

Those whose works the author has used in his work and first and foremost V.M. MAIDEBOR.

Max CREUSOT who has used all his talent in translating this work into the English language.

And last, but not least, those whose names are not cited, no doubt because their works are already so much a part of our savoir-faire such as M. MUSKAT and S.J. PIRSON.

L.H. REISS

contents

symbols

a = typical size of a matrix element

A_1, A_2, A_3 = coefficients used by Pollard for well test analysis

A = coefficient of the well performance equation $\Delta P = AQ + BQ^2$

b = typical width of a fracture

B_o, B_w, B_g = oil, water and gas formation volume factors

B = coefficient of the well performance equation $\Delta P = AQ + BQ^2$

C = compressibility

D = diffusivity

E_i = error function

E_t = recovery factor

f_s = cumulative length of fracture per unit cross section perpendicular to the direction of flow

F = front

g = acceleration due to gravity

H = distance between the completion interval and the fluid contact for coning equations

PI = productivity index

k = permeability

h = thickness

l = width

L = length

m = slope of a pressure build-up or drowdown plot on semi log paper

M = point

n = number of fracture planes

N = oil in place

P = pressure

q = flow rate

r = radius

s = skin effect

S = saturation

S_{cw} = connate water saturation

S_{or} = residual oil saturation

t = time

T = temperature

x, X = coordinates in space

α = ratio of permeability from well tests to core permeability

$\alpha_1, \alpha_2, \alpha_3$ = coefficients used by Pollard for well test analysis

β = coefficient of thermal expansion

$\Sigma_1, \Sigma_2, \Sigma_3$ = coefficients

θ = wetting angle

λ = experimental factor for turbulent flow

μ = viscosity

π = dimensionless group or the constant pi

ρ = specific gravity

σ = surface tension

ϕ = porosity

Subscripts

c = capillary (pressure, etc.)

b = bubble point

cr = critical

d = threshold

Symbols

D	= dimensionless		p	= pore or produced
e	= effective or external		r	= relative or residual
f	= fracture or flowing		s	= surface
g	= gas		si	= shut in
i	= initial		t	= total
m	= matrix		vs	= secondary porosity
o	= oil		w	= water or well

1

introduction

Fractured reservoirs were deposited as conventional sediments of the matrix type, with intergranular porosity: their continuity has been disrupted as a result of tectonic activity. These discontinuities introduce considerable difficulties in the description of both the internal structure and the flow of fluids within fractured reservoirs. This additional complexity is often compounded by diagenesis which is strongly affected by the preferential flow of waters through the fracture network: deposition of minerals such as calcite, dissolution of the matrix sometimes leading to very large cavities. The formation of discontinuous chemical residues (stylolites) is often associated with the compressive components of the tectonic stresses involved in fracturing.

Thus in addition to the conventional parameters that are necessary to describe the matrix – e.g. permeability, porosity – the production geologist has to evaluate three discontinuous networks of:

(a) Fractures.
(b) Open channels within the fracture system.
(c) Stylolites.

to which may often be added the presence of vugs. The fracture system dominates the flow of fluids in such reservoirs, and a description of these interconnecting networks is an essential prerequisite to a reservoir engineering evaluation.

The production geologist's main problem lies in the type of data which is available to him, which is basically unsuitable. Extrapolations and correlations which are routine for conventional reservoirs are not valid when evaluating discontinuities whose spacing is roughly the same size as the width of the sample available – the core diameter – especially when these samples are taken at widely spaced intervals as when development (let alone exploration) drilling is carried out.

Thus even more so than with conventional formations, it is necessary to inte-

grate several disciplines to evaluate fractured reservoirs. Use is made of sedimentological techniques, of the tectonic history of the field, of mathematical
models of rock mechanics, of production data and even in some cases of seismic
sections to improve the initial geological description made from cores.

Production tests are most significant: we shall see that the network of open
channels through which flow takes place can rarely be defined from direct
observation, because the fracture parameters (especially width) cannot be measured with any accuracy. It follows that production tests are the only reliable
means of estimating the flow characteristics and productivities of fractured
reservoirs.

Most fractured reservoirs correspond to the simple scheme of matrix elements
separated by fractures illustrated in Fig. 1. The original matrix is usually not
very permeable — the compactness of porous rock is directly related to its
rigidity and tendency to fracture — but its porosity can vary from high values
(such as in chalks) to very low values (the case of karsts). The fracture network
is often surprisingly regular, as shown by the photographs of outcrops in
Fig. 9, so that the reservoir may be thought of as consisting of elements of low
permeability matrix separated from each other by fractures which may be
closed and filled with cement, or may still act as effective channels for flow.
According to Davadant (Ref. 22) these channels range from about ten microns
upwards, although the lower values are most commonly encountered. The
matrix elements can become increasingly vuggy towards the fractures.

Capillary phenomenon, and in particular connate water saturations, can
usually be ignored when dealing with fractures. Smekhov (Ref. 35) has shown
that the film of water which adheres to the sides of fractures as a result of
molecular forces is at most 0.016 microns (see Fig. 2). This implies that the two
films on opposite faces of fractures at least 10 microns wide never meet to
form a meniscus, and that capillarity plays little or no role in the fracture
network. The conventional behaviour of the matrix, however, is in no way
affected.

We shall use the terminology adopted by Maidebor (Ref. 1) and distinguish
porous from non porous fractured reservoirs (see Fig. 3) :

(a) Porous fractured reservoirs are the most common: examples are the
Iranian fields, Parentis (France), Rhourde El Baguel (Algeria), Ekofisk (North
Sea), Sprawberry (USA).

For practical purposes all of the oil in place is within the matrix, and flow
towards the wells takes place in the fracture network.

(b) Non porous fractured reservoirs are dominated by karsts such as found
in the USSR, Nagylendel (Hungary), Rospo (Italy), Emposta (Spain).

The matrix is impermeable and contains no oil in place: reserves and fluid
flow are restricted to the network of open channels and connected vugs.

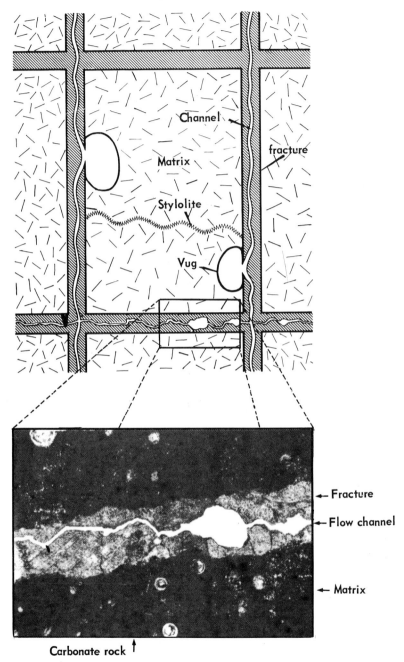

Fig. 1.   Idealized matrix element and illustration of secondary
porosity within a fracture.

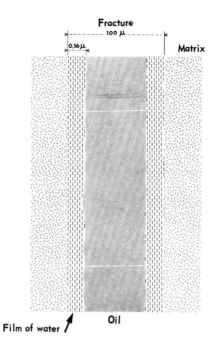

Fig. 2.   Illustration of oil and water distribution in a fracture at
equilibrium.

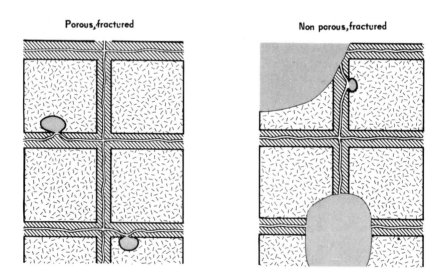

Fig. 3.   Illustration of porous and non porous, fractured reservoirs.

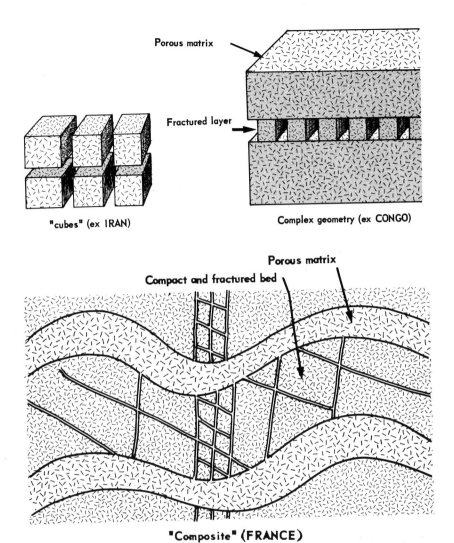

"cubes" (ex IRAN)　　　　　Complex geometry (ex CONGO)

"Composite" (FRANCE)

Fig. 4.　Examples of fractured reservoirs.

This definition is unfortunately ambiguous, because non porous fractured reservoirs often have a small matrix pore volume filled with water: the porosity plays no role in fluid flow, and a better terminology, unfortunately not in common usage, would have been "impermeable fractured reservoirs". Even so, there remains the intermediate case of the very tight matrix whose pore size is so small that oil has not been able to displace water from the pores (Maria Mare, Italy), but whose connate water can be expelled during depletion. These reservoirs are normally classed among the non porous fractured reservoirs.

This division of fractured formations according to the matrix properties can be extented to take into account:

(a) The number of fracture planes: three orthogonal planes will divide the reservoir into cubic elements such as in Iran, two planes will lead to an arrangement of "match-sticks", one plane will lead to sheets.

(b) The intensity and regularity of fracturing (see Fig. 4): in many reservoirs, alternating layers of permeable and tight rock have reacted differently to tectonic stress, giving a system of fractured beds sandwiched between unfractured beds (Emeraude, offshore Congo). Lateral changes in facies can lead to reservoirs which are locally fractured. The tectonic stress is not uniform and the intensity of fracturing may be more pronounced near tectonic features such as strong folding or faults.

In the following chapters we shall discuss the two objectives facing the petroleum engineer evaluating a fractured reservoir: the geological description, and the flow of fluids. This separation is artificial but inevitable: in practice the two aspects are interconnected and considerable feed back is necessary between the geologist and the reservoir engineer.

2

the production geology
of fractured reservoirs

The presence of fractures does not affect the description of the matrix which will not be dealt with as it does not differ from conventional reservoir geology. We shall be concerned with the discontinuous networks which distinguish fractured formations.

Production geology is a descriptive technique ; in fractured reservoirs it is based on cores: cuttings and side wall samples are never used for obvious reasons of physical size. The importance of coring fractured reservoirs cannot be over emphasized.

In many cases core recovery is very poor; drilling fractured formations is usually associated with mud losses, and the weakness of the fracture plane can lead to mechanical failure of the rock as it is being cored. Nevertheless, however discouraging, an attempt must be made to adjust drilling and coring techniques to the reservoir being penetrated because it is the only direct method of observing the fracture network and obtaining information which we shall see can condition such fondamental choices as gas or water injection.

We shall include a discussion of the use of outcrops and rock mechanics in this chapter; electric and nuclear logs have not yet been interpreted successfully in terms of fracture network and are discussed in Appendix 2.

2.1. DESCRIPTION OF CORES

A schematic view of a typical core taken from a fractured reservoir is shown in Fig. 5.

Great care has to be taken when describing fractures. When traces of relative movement either side of a fracture can be seen, the break may be due to the mechanical action involved in coring. Breaks occurring parallel to the bedding

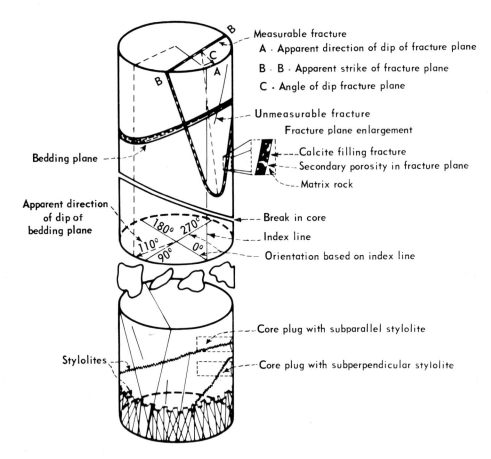

Measurable fracture
A . Apparent direction of dip of fracture plane
B . B . Apparent strike of fracture plane
C . Angle of dip fracture plane

Unmeasurable fracture
Fracture plane enlargement
Calcite filling fracture
Secondary porosity in fracture plane
Matrix rock

Break in core
Index line
Orientation based on index line

Core plug with subparallel stylolite
Core plug with subperpendicular stylolite

Bedding plane

Apparent direction of dip of bedding plane

Stylolites

Fig. 5.   Definition of the parameters used to describe cores.

are excluded because they are usually due to handling. The following parameters are used to describe fractures — a typical data sheet is shown in Fig. 6:

(a) Distance between fractures.
(b) Dip and direction of the fracture plane.
(c) Width, degree of cementation, length.

In practice, width is often too small to be measured, so that only exceptional values are recorded: this implies that fracture porosity is rarely estimated from core descriptions. Estimation of dip (in deviated wells) and direction of the fracture planes requires oriented cores.

Stylolites form separations within the matrix (see Fig. 5): they consist of complex chemical residues due to reactions under temperature and pressure. Their shape is irregular and dented, the thickness of the stylolite itself being a

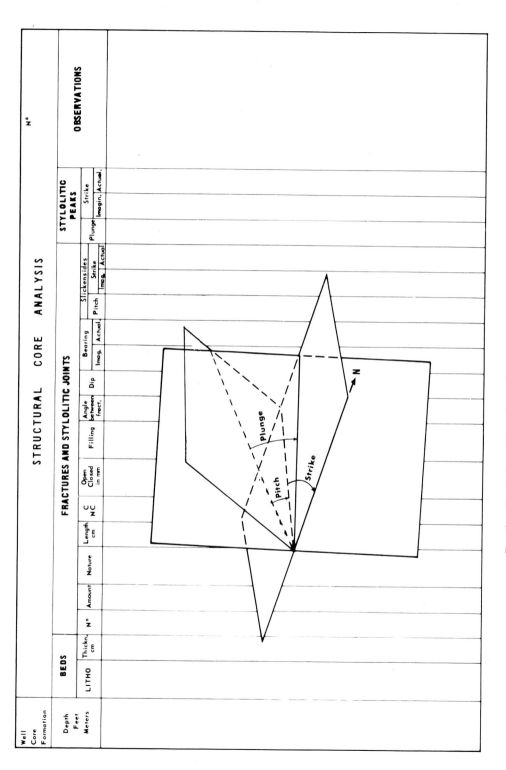

Fig. 6. Typical core data sheet. (After Jean et Masse, Ref. 2).

few millimetres although the amplitude between its peaks can be several centimetres. Their amplitude, thickness and frequency is recorded.

Core descriptions are usually computer processed because of the quantity of
information available for analysis. The treatment is statistical and inspired by
the results obtained from the study of outcrops and rock mechanics. We shall
discuss two examples of this type of evaluation:

(a) The main directions in which fracturing takes place can be shown on a
stereographic diagram (see Fig. 7). The orientation and dip of each fracture is
indicated on a circular chart graduated in dip and direction on scales marked off
respectively on the radius and periphery: groups of points indicate the trend
of the main fracture system — in the case of Fig. 7, two orthogonal directions.

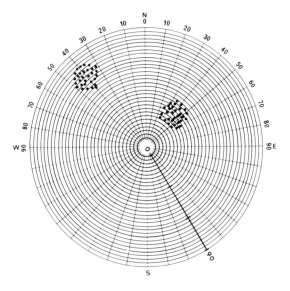

Fig. 7.   Stereographic diagram.

(b) The size of the matrix elements can be displayed by a diagram explained
in Fig. 8: the fracture is assimilated to a plane infinite up and downdip, but
limited to the width of its intersection with the core. The intersection of this
fracture plane with another plane —usually vertical— gives a trace of the fracture. A combination of all the traces due to the fractures on a core gives a good
idea of the intensity of facturing in the reservoir, and can be used to estimate the
average size of an element of matrix. Fig. 8 is taken from a field case and the
comparison of this visual display based on core data with the photographs taken
from outcrops such as given in Fig. 9 shows what a realistic image of the
reservoir can be obtained from such an analysis.

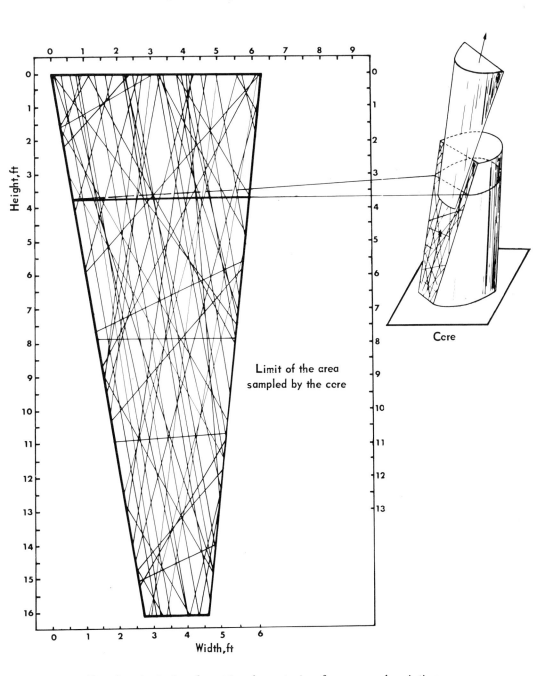

Fig. 8.   Analysis of matrix element size from core description.
(After *Corelab*; see also Ref. 5)

Fig. 9.   Some examples of blocks (Ref. 3 and internal Elf documents,
photos of outcrops).

2.2.   DESCRIPTION OF OUTCROPS

The study of outcrops is the best way to obtain an overall view of the relationship between tectonics and fractures, stylolites, channels, etc. The finer details can only be seen on fresh outcrops such as are found in quarries and recent road cuttings, but fairly large outcrops are required if statistical treatment is to have any meaning. The measurements which must be made are much the same as in the case of cores, to which is added a description of the deformation of the beds affected by fracturing: an attempt is made to correlate the degree of deformation with the onset of fracturing, the obvious application being the inference of the fracture intensity from structural maps. Figure 9 illustrates some typical features of the outcrops of highly fractured reservoirs.

2.3.   APPLICATION OF ROCK MECHANICS

Rock mechanics involve the calculation of the stress within a reservoir subjected to tectonic constraints, and its application is the prediction of fractures from a knowledge of the rupture point of the rock. The stress in porous rock is the difference between "mechanical" forces such as the weight of the overburden, and the fluid pressure: this is an important concept which will be encountered several times, for instance when discussing the decline in well productivity during depletion.

The four elements involved in mathematical models of rock mechanics are:

(a) The weight of the overburden, which can be calculated from data such as density logs and formation thickness.

(b) The fluid pressure, which can be measured directly.

(c) The elastic properties of the rock (Young's modulus, Poisson's ratio, the rupture point) which can be determined experimentally in the laboratory.

(d) The tectonic forces acting on the strata which are calculated by a mathematical model using structural maps based on seismic profiles, one close to the reservoir and another well above it.

Many mathematical models of rock mechanics have been written: they are usually based on the theory of elasticity for homogeneous isotropic media, but finite element techniques have led to models which can handle more complex situations. They do not represent the discontinuity due to fracturing, when the rock properties cease to be elastic, but give a map of the stresses involved in the reservoir if no fracturing had occurred from which probability distributions of the intensity of fracturing can be derived. Their main application is there-

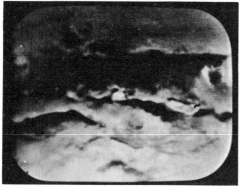

281.60 m (924 ft)
Horizontal joint stuffings

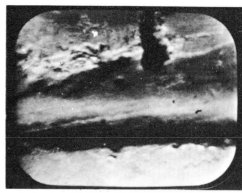

276.60 m (907 ft)
Vertical fracture interrupted by a silt bank

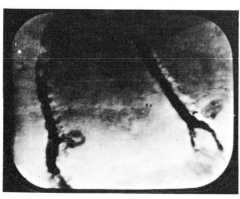

276.65 m (907 ft)
Fractured limestone bank

276.65 m (907 ft)
Same fracture but on
opposite side of bore hole

275.20 m (903 ft)
Vertical fracture interrupted
by a facies change

237.70 m (779 ft)
Crushed zone or sedimentary breccia

Fig. 10.   Photographs taken on television screen in Emeraude (offshore (Congo)) (Magnification approx. 1).

fore as a help in extrapolating well data to parts of the reservoir where different stresses have occurred. There is an inconsistency in this type of model which has unfortunately not yet been resolved: the presence of the first fractures modifies the stress in the reservoir, and its incidence on later fracturing is not taken into account.

2.4. OBSERVATION IN SITU

Several tools have been recently developed so that the well bore can be observed directly: they are discussed with other logging techniques in Appendix 1.

The "Bore Hole Tele Viewer" (BHTV) operates on the same principle as the sonar: its use is not restricted by practical considerations, but it gives a poor image and is seldom used.

Photography and closed circuit television (Ref. 4) have given excellent results — an example from the Emeraude field, offshore Congo, is shown in Fig. 10. They can only be used in exceptional circumstances: being optical tools, they require the absence of mud cake and a transparent fluid in the well.

2.5. USE OF SEISMIC PROFILES

Intense fracturing can disturb the acoustic properties of a formation, and this may lead to anomalies on seismic profiles. Fig. 11 is taken from the Emeraude field (offshore Congo): once the reservoir was known to be highly fractured, all the seismic profiles were compared so as to give an areal view of the intensity of the seismic anomaly, which can be correlated with the reservoir properties (Ref. 38).

2.6. CONCLUSION

Production geology data is treated statistically: the objectives are:

(a) The shape and average size of the matrix elements.
(b) The intensity of fracturing throughout the field.

The results obtained are approximate and often inconclusive because of the dimensions of the samples available — cores — as compared to the frequency of the discontinuities being evaluated — fractures —, and because of the assumptions inherent to the analytical tools available.

The other parameters defining a fractured reservoir, particularly those which define fluid flow such as fracture width, have to be evaluated using other techniques. In the following chapter, the contribution of production tests to the understanding of fractured reservoirs will be discussed.

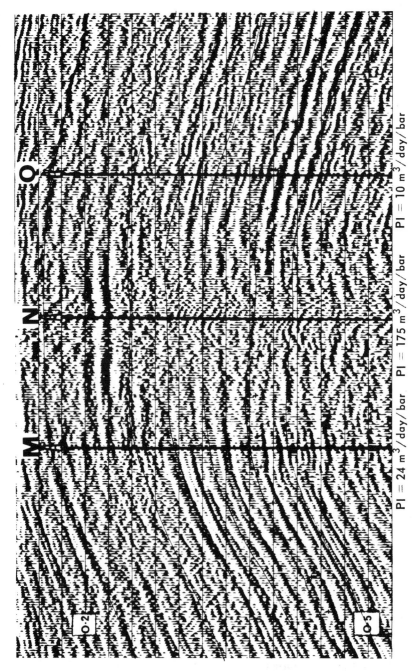

PI = 24 m³/day/bar   PI = 175 m³/day/bar   PI = 10 m³/day/bar

Fig. 11.   Illustration of the use of seismic profiles to detect fractures taken from the Emeraude field (Congo). The profiles are clear on the flanks (wells M and Q) but disturbed at the centre (well N) which is highly fractured and has an exceptional productivity.

3

the use of production data
in fractured reservoirs

In this chapter we shall be concerned with the information that can be derived from the analysis of well behaviour. A detailed discussion of the topic will be found in Appendix 2. Although several attempts have been made to establish interpretation methods which will give the fracture parameters, in practice well tests are usually analysed in the same way as for conventional reservoirs.

The first point to be made is that flow near the well bore is often turbulent, so that the productivity index may not be constant, and well performance predictions usually require several periods of stabilized flow at different rates.

Pressure drawdowns and build-ups can usually be interpreted to give the fracture permeability. A parameter that is often used to reflect the intensity of fracturing is:

$$\alpha = \frac{\text{permeability from drawdown or build-up}}{\text{matrix permeability from cores}}$$

α can range from about ten when few channels are open to flow, to values of several thousand or more in the case of highly fractured reservoirs.

As production tests are usually carried out over relatively thick intervals, production logs are required to distinguish the fractured zones responsible for most of the production: both flow-meters and temperature surveys are used.

When enough wells are available, an attempt can be made to correlate parameters such as the intensity of fracturing throughout the field: this type of analysis often leads to obvious results such as high productivities near faults, but can also suggest features that were not previously suspected, especially when the seismic profiles are of poor quality or widely spaced.

Interference tests are of interest in both porous and non porous fractured reservoirs for two reasons: they give an estimate of the continuity of the fracture system, and help detect anisotropy. They are of major importance in the case of

non porous fractured reservoirs as a method of estimating the secondary porosity, which is difficult to establish, and which, in this case, means oil in place
since the matrix is not oil bearing.

The feed back between production geology and well test results is usually
done through the relationship which exists between the fracture parameters:

k_f = fracture permeability,
ϕ_f = fracture porosity,
b = fracture width,
a = typical dimension of the element of matrix.

These relationships are discussed in more detail in Appendix 3. They are
based on simple geometrical systems such as shown in Fig. 12 which are idealized representations of cases frequently encountered by geologists as will be

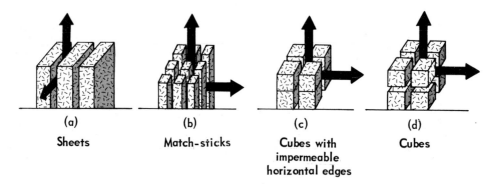

(a)	(b)	(c)	(d)
Sheets	Match-sticks	Cubes with impermeable horizontal edges	Cubes

Fig. 12.   Illustration of matrix element shapes.

immediately apparent from the regularity of the features photographed on
outcrops such as those shown in Fig. 9. Figure 13 is a typical example of a
theoretical relationship between k, ϕ, b and a corresponding to cubic matrix
elements of side a — the situation depicted in Fig. 12 d. Knowing k from production tests and a range of values of b leads to both a and ϕ. A cross-check with
core descriptions can then be made.

The following equation is also used to give a minimum value of ϕ_f:

$$\phi_f = \frac{1}{577.9} \left(\frac{PI \, \mu_o \, B_o \, f_s^2 \, \log \frac{r_e}{r_w}}{h} \right)^{1/3} \tag{1}$$

where

f_s is a parameter in cm/cm^2 discussed in Appendix 3 and,
PI = productivity index in (m^3/d/bar),
μ_o = viscosity (cPo),

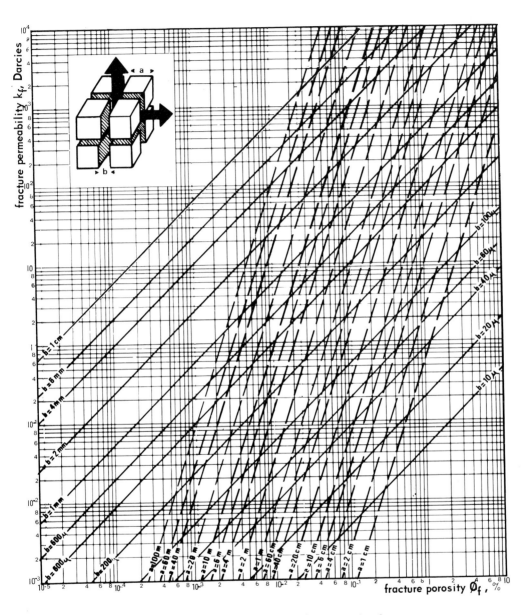

Fig. 13.   Example of the relationship between the fracture parameters for cubic matrix elements.

B_o = formation volume factor (dimensionless),
r_e = drainage radius (cm),
r_w = well radius (cm),
h = formation thickness (m).

This equation is Darcy's law for radial flow in a fractured reservoir consisting of impermeable cubic elements.

In the case of waterfloods of non porous fractured reservoirs, an estimate of fracture and vug porosity can be made from observations of the moving oil water contact: unfortunately this can only be done after considerable reservoir history is available.

Production data in its widest sense includes reservoir performance. In the absence of water drive, the pressure against cumulative production decline curves can be used to estimate the fracture pore compressibility C_{pf}: it is usually slightly larger but of the same order of magnitude as the matrix pore compressibility C_{pm}, 10^{-5} bar^{-1} (Appendix 4).

Well behaviour is affected by the presence of fractures, and in particular the productivities and injectivities depend on the state of depletion of the reservoir. Productivity decline is attributed to the gradual closing of the fractures as the fluid pressure is reduced while the mechanical constraints remain unchanged. There is an approximate relationship (Appendix 3.4):

$$\frac{k}{k_{initial}} = (1 - C_{pf} \, \Delta P)^3 \tag{2}$$

where

k represents permeability (mD),
ΔP = pressure drop (bar),
C_{pf} = fracture pore compressibility (bar^{-1}).

Taking the range 10^{-4} to 10^{-3} bar^{-1} for C_{pf} and a 50 bar pressure drop gives a 1.5 to 15% decline in productivity. These values may appear at first glance to be small. The explanation is thought to lie in the presence of secondary cement within the fracture system which helps to resist the increased effective pressure to which it is subjected, and maintains the network of flow channels open.

Injectivities are often higher than productivities (Appendix 2) and while an increase in fracture width due to pressure may be partly responsible, the main cause is probably the shrinking of the matrix due to the drop in temperature associated with the injection of colder fluids.

At this stage of reservoir evaluation, an estimate of the following parameters should be available:

(a) Block size = a few centimetres to several metres.

(b) Fracture width = ten microns upwards, values as high as a few millimetres are usually associated with exceptional productivities.

(c) Fracture permeability = as low as ten millidarcies corresponding to exceptionally narrow fractures: permeabilities above one darcy are common, and as a general rule fractured reservoirs are much more permeable than conventional ones: one hundred darcies has been measured on Emeraude (Congo) where the 100 cP oil makes pressure build-ups amenable to interpretation.

(d) Fracture porosity = between one part per thousand and one per cent.

(e) Vugular porosity for karsts = about one per cent.

The range of values given are only orders of magnitude and many exceptions can be found: for instance, some shallow reservoirs with very wide fractures have a fracture porosity of several per cent. Fracture widths of the order of a millimetre are restricted to exceptional fields such as those in Iran: values below 100 microns are more usual and provide adequate permeabilities for commercial production.

4

recovery mechanisms
in fractured reservoirs

4.1. INTRODUCTION

The specific problem of non porous fractured reservoirs will be dealt with as an exception: we shall consider the typical but more complex case of porous fractured reservoirs as the rule.

The same physical principles control recovery from both fractured and conventional reservoirs: the difference lies in the relative importance of the quantities involved.

If there were no open channels, very few fractured reservoirs could be of commercial interest because the low matrix permeability would lead to uneconomic well productivities. The basic role of the fractures is to act as a connection between the oil bearing matrix and the wells. The combination of high porosity/low permeability matrix and low porosity/high permeability fracture is the key to the performance of porous fractured reservoirs.

The recovery mechanisms are basically fluid expansion, pore volume contraction, displacement of oil from the matrix, convection and diffusion.

4.2. EXPANSION

Figure 14 shows the basic element of a fractured reservoir: matrix, stylolites, vugs surrounded by a fracture system within which there is a network of flow channels.

We shall ignore the presence of stylolites. The few measurements made on cores indicate that they are not absolute barriers to flow, but this cannot be

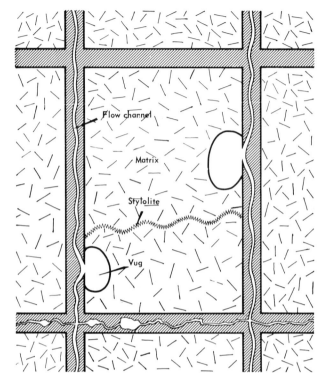

Fig. 14.   Illustration of an element of a fractured reservoir.

regarded as conclusive because stylolites are fragile and may have been broken during coring or the preparation of plugs for laboratory experiments. Moreover stylolites divide the elements of matrix but these elements are still in touch with fractures on either side, and it is extremely rare that the network of stylolites completely surrounds a portion of matrix and isolates it from the flow channels. Thus even if stylolites were impermeable, the matrix would still be in communication with the fractures although the geometry of the connection would be slightly more complex.

During single-phase expansion, the material balance can be written (AIME symbols):

$$NC_{et} \, \Delta P = N_p \tag{3}$$

where C_{et} is the total effective compressibility to oil of the system and includes rock, fractures, oil and water. In Appendix 4, C_{et} is shown to be:

$$C_{et} = C_o + \frac{(C_w S_{wm} + C_{pm}) \, \phi_m + C_{pf}\phi_f}{\phi_m \, (1 - S_{wm}) + \phi_f} \tag{4}$$

where

C_o = oil compressibility,
C_w = water compressibility,
C_{pm} = matrix pore compressibility,
C_{pf} = fracture pore compressibility,
ϕ_m = matrix porosity,
ϕ_f = fracture porosity,
S_{wm} = initial matrix water saturation.

In the case of porous reservoirs this equation can be simplified: ϕ_f being small as compared to ϕ_m, and C_{pf} and C_{pm} being of similar magnitude:

$$C_{et} = C_o + \frac{C_w S_{wm} + C_{pm}}{1 - S_{wm}} \tag{5}$$

Thus expansion is almost entirely due to the properties of the matrix and its fluid. The presence of a network of flow channels implies that this expansion is uniform throughout the matrix. An estimate of the duration of the transient period Δt before semi-steady state prevails (and the rate of pressure decline is identical at the centre of the matrix element and on its periphery at the fracture) is mentioned by Dupuy (Ref. 6) who quotes Chatas:

$$\Delta t = 500 \, \frac{a^2 \, \phi_m \, \mu \, C_{pm}}{k_m} \quad \text{second} \tag{6}$$

Typical values would be:

a = linear dimension (100 cm),
μ = viscosity (1 cP),
ϕ_m = matrix porosity (0.10),
C_{pm} = matrix pore compressibility (1×10^{-4} bar $^{-1}$),
k_m = matrix permeability (1 mD).

giving $\Delta t = 50$ s. It follows that for practical purposes the matrix element and its surrounding fractures decline in pressure at identical rates and that the matrix pressure lags behind by at most a few minutes.

Fractures therefore play a beneficial role during single phase depletion: the pressure decline is more uniform than is the case for conventional reservoirs. One consequence is that as long as the fracture network provides adequate drainage all the oil in place is active and no permeability or porosity cut off (such as is commonly used for conventional reservoirs) is required when estimating the oil in place active during the depletion of a fractured reservoir.

These conclusions also hold when reservoir pressure declines below bubble point: the total compressibility is increased by the presence of gas, and fracture compressibility has even less influence. In addition when its saturation reaches the critical value for flow, the liberated gas bleeds-off into the flow channels where it migrates under the influence of gravity towards the crest of the reservoir to form a secondary gas cap.

Solution gas drive in fractured reservoirs differs in some ways from the same mechanism in conventional reservoirs. It is known (Ref. 39) that the formation and growth of gas bubbles is encouraged by a homogeneous permeable porous network of low porosity. As a result the gas bubbles tend to form and coalesce in the fractures. The gas in the fractures then absorbs the lighter molecules in the matrix by diffusion and this process reduces the impact of solution gas drive as a mechanism for expelling oil from the matrix. A case is cited in Ref. 8 of an Iranian reservoir whose pressure was lowered 500 psi below bubble point and whose free gas saturation in the matrix did not exceed 1%. This subject is also mentioned in Section 4.3.

4.3. SUDATION

The idealized element of a fractured reservoir is illustrated in Fig. 15 : matrix saturated with oil and partially or completely submerged in water or gas. By the

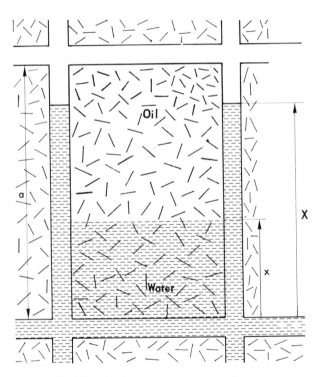

Fig. 15.   Sudation from a matrix element.

term "sudation" we refer to the combined effects of two sets of forces which play a role in the substitution of oil within the matrix by the water or gas in the surrounding fractures:

(a) Gravity forces due to the difference in densities between oil and water (or gas).

(b) Capillary forces due to the interaction of surface forces within the pores.

Sudation is discussed in detail in Appendix 5. A large number of assumptions need to be made so that the physical process can be described mathematically, and we shall see that in practice laboratory experiments, even though they are questionable, are often preferred to theoretical results. Nevertheless we shall use the theoretical equations as a basis for the qualitative description of a particular case which will illustrate the interplay of gravity and capillarity. The initial rate of expulsion of oil q_i per unit cross section from an element of matrix suddenly completely immersed in water can be derived from the equation presented in Section 2 of Appendix 5:

$$q_i = \frac{k_o}{\mu_o} \frac{a(\rho_w - \rho_0)g + P_c}{a} \tag{7}$$

where

k_o . = matrix permeability to oil,
$\rho_{w,o}$ = water and oil specific gravities,
μ_o = oil viscosity,
a = typical **vertical** dimension of the matrix element,
P_c = capillary pressure.

The term $a(\rho_w - \rho_o)g$ represents the magnitude of the gravity forces, and is proportional to the dimensions of the matrix. The necessary condition for oil expulsion to take place is $q_i > 0$:

$$a(\rho_w - \rho_o)g + P_c > 0 \tag{8}$$

We shall discuss the different cases separately:

Oil and water, water-wet matrix.

Water has a natural tendency to penetrate the matrix, and gravity reinforces capillary imbibition: both terms of Eq. (8) are positive and $q_i > 0$, i.e. oil is displaced by water.

When gravity becomes negligeable, as is the case for small matrix elements, the process becomes, for practical purposes, capillary imbibition.

Oil and water, oil-wet matrix.

For an oil-wet matrix, capillary forces oppose the penetration of water into the matrix, and displacement is only possible if the driving force (gravity) overcomes the resistance which we shall call the threshold capillary pressure P_d.
Displacement only if:

$$a(\rho_w - \rho_o)g > P_d \qquad (9)$$

Note that this is only possible for matrix elements of a certain size (a large). It follows that oil cannot be expelled by water from an intensely fractured oil-wet reservoir.

Oil and gas.

Capillary pressure in a water/oil system is always given as the difference between the pressure in oil less the pressure in water, and is positive for water-wet rock. Reservoir rock is always oil rather than gas-wet, and capillary pressure in the oil gas system is defined as the pressure in gas less the pressure in oil; this definition involves a change in sign of Eq. 7 which becomes:

$$q_i = \frac{k_{ro}}{\mu_o} \; \frac{a(\rho_o - \rho_g)g - P_c}{a} \qquad (10)$$

where

ρ_g is the gas specific gravity.

By analogy with the oil-wet rock in the oil/water system discussed above, displacement only if:

$$a(\rho_o - \rho_g)g > P_d \qquad (11)$$

An extensive discussion of recovery mechanisms will be found in Ref. 7.
In the case of water-wet rock, the best recoveries are often obtained from waterfloods because both gravity and capillarity combine to expel oil from the matrix. However in some cases the larger gravity term in the case of a gas/oil rather than water/oil system can compensate the reversal in sign of the capillary pressure, so that gas injection is more favourable. The choice of which fluid to inject depends on the matrix dimension a, and illustrates the importance of a good geological description of fractured reservoirs when planning their development.
The case of oil-wet reservoirs has recently gained emphasis as a result of the failure of a number of waterfloods carried out on limestones: it would appear that the presence of small quantities of organic matter (such as coal) dispersed in the matrix can induce a wettability to oil which results in poor reservoir performance under water injection.

Sudation may sometimes be impossible: this is the case of intensely fractured reservoirs with matrix elements of a few centimetres surrounded by a secondary gas cap or by water if the matrix is oil-wet.

Sudation is usually described by "transfer functions" often misleadingly referred to as "imbibition curves" (Appendix 5). These curves are simply the quantity of oil expelled from a matrix element surrounded by fractures as a function of time. In the water/oil system they are usually based on laboratory work, but gas/oil transfer functions are normally derived mathematically because of the experimental difficulties involved in gas flooding at reservoir pressure and temperature cores of the size of matrix elements commonly encountered [1]. "Imbibition curves" are discussed in Appendix 5.

Each case must be evaluated on its own merits, and laboratory work or mathematical simulation are an absolute prerequisite to the choice between gas (including secondary gas cap) and water (including aquifer) drives.

There are two phenomena also worth mentioning in connection with the sudation process: block to block interaction and the influence of "bridges".

(a) Block to block interaction or cascade effect which can occur when the matrix is oil bearing and the fractures filled with gas: the oil droplets expelled by gravity dominated sudation at the top of the reservoir can be reabsorbed by the matrix on their journey downwards towards the GOC in the fracture network by capillarity.

(b) Bridges. The matrix blocks can be interconnected (in a capillary network sense): there exists matrix "bridges" between them. This may in certain cases improve considerably the sudation process in comparison with the case of completely isolated blocks by increasing the effective element size a.

Both phenomena are mentioned in Ref. 17 ; the cascade effect is studied in some detail in Ref. 42.

4.4. CONVECTION AND DIFFUSION

Convection and diffusion are often ignored when dealing with conventional reservoirs, because of the very large time scales required before their effects become significant. The presence of a network of high permeability channels accelerates these phenomena which have been detected in thick, highly fractured oil pools.

------

[1] Refer also to the comments made in the following chapter on the subject of "diffusion".

Convection (Refs. 8, 9)

Convection is the result of instability due to the presence of oil at the crest of the reservoir which is heavier than towards the base: the vertical fractures in thick reservoirs provide the communication for convection to take place so that equilibrium is reestablished.

Instability prevails if:

$$F = \beta_o \frac{dT}{dz} - C_o \frac{dP}{dz} > 0 \tag{12}$$

An example taken from Iran (Ref. 8) gives:

β_o = coefficient of thermal expansion at constant pressure $(1.15 \times 10^{-3} v/v/°C)$,

$\dfrac{dT}{dz}$ = variation of temperature with depth $(3.64 \times 10^{-2} °C/m)$,

C_o = compressibility at constant temperature $(1.02 \times 10^{-4} v/v/bar)$,

$\dfrac{dP}{dz}$ = variation of pressure with depth $(0.069\ bar/m)$.

leading to $F = 3.49 \times 10^{-5} > 0$ so that reservoir oil is not in equilibrium.

A consequence of the existence of convection currents is that important variations of bubble point with depth are rare: this is not the case for conventional reservoirs where gradients of saturation pressure are commonly encountered.

Diffusion

Convection is a result of contrasting oils within the fracture network. Diffusion is due to the contrast in hydrocarbon properties between fracture and matrix: it can take place between gas and oil, thus enchancing sudation, or between oils with different compositions.

This phenomenon has been observed in several Iranian fields (Ref. 8) where the saturation pressure has been found to change by as much as 35 bars during ten years of single-phase depletion.

Diffusion can be studied by means of suitable mathematical models (see C-FRAC and YAMAMOTO models in Appendix 6). Furthermore, some laboratories (Ref. 43) are today equipped with physical models, which enable experiments to be performed on rock samples several metres long. These experiments particularly concern the exchange of components between the gas in the fissures and the oil in the matrix, under bottom-hole conditions and over the actual period of time this takes (several months). This type of experiment can of course be used to calibrate mathematical models.

4.5.   MULTIPHASE FLOW IN THE FRACTURE NETWORK

For simplicity we shall restrict ourselves to waterfloods: gas drive leads to similar results with the exception of diffusion which can lead to the enrichment of the injected gas with light and intermediate components of the oil.

The flood front moves more rapidly in the fracture network, and by-passes the oil in the matrix: sudation intervenes, and some of the matrix oil is interchanged with water from the fractures, so that an oil/water mixture flows in the network of channels. The oil in the fractures migrates upwards under the influence of gravity. This multiphase flow is described using the relative permeability concept.

Laboratory work (Ref. 10) has shown that:

(a) For a single fracture, the fluids have a strong tendency to interfere with each other and the relative permeabilities reflect a decrease in total mobility.

(b) For a connected network of fractures, the fluids segregate and flow in different channels: overall mobility is not affected.

(c) Segregation takes place rapidly as compared to the time scale involved in field operations.

The relative permeabilities used for flow in the fracture network are therefore straight lines, and the absence of capillary phenomenon in the fractures means that their end points at water saturations of 0 and 100 % are both close to one. It is common practice to introduce 20 % residual oil saturation, but this is designed to take into account the oil trapped as a result of the irregularities of the flow channels (which may be completely cemented in places) and not the physical process at work.

It is usually assumed that flow in the fracture network is always single phase: either oil, water or gas. This leads to the concept of a water or gas level in the fracture network. Because of the low fracture porosity, the volumes ignored such as gas migrating towards a secondary gas cap are negligeable, and a result of the high permeabilities of the fracture network is that the fluid levels are often nearly horizontal (see Section 4.9).

4.6.   RELATIVE MOVEMENT OF THE OIL/WATER CONTACTS

The water level in the matrix lags behind the water level in the fractures. The concept of a critical rate at which the water rises at the same speed in both fractures and matrix has been put forward. If withdrawals exceed the critical speed the water in the matrix lags more and more behind. Laboratory studies have shown that more oil is recovered by sudation into a partially submerged element of matrix than from one that is completely submerged. Unfortunately in most cases economic necessity dictates the rate of withdrawals and the reservoir engineer has to design his laboratory experiments accordingly. In some rare cases, it is possible to adjust the production rate to optimize this aspect of recovery.

4.7.   INTERPLAY OF THE RECOVERY PROCESSES

We have seen that fractures play a positive role during single-phase depletion which is often important because the aquifers of fractured reservoirs are frequently inactive — for instance when the fracture network is filled with secondary cement such as calcite or when fracturing is less intense away from the structure. Recovery by single-phase expansion can be as high as 10% for undersaturated oils such as Rhourde el Baguel (Algeria) and Ekofisk (North Sea).

When pressure declines below bubble point, gas is liberated and as lateral pressure gradients towards the wells are small due to high fracture permeabilities, the gas migrates upwards under the influence of gravity to form a secondary gas cap. This has two results: the rate of pressure decline decreases because of the presence of more compressible gas; and sudation of oil from the matrix takes place at the crest of the reservoir. Sudation may also occur simultaneously between water (which has its origin in the aquifer) and the oil at the base of the reservoir. Figure 16 illustrates the different recovery processes which occur

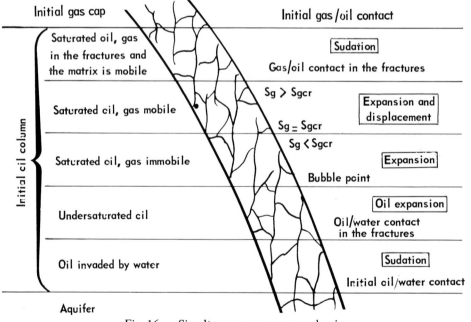

Fig. 16.   Simultaneous recovery mechanisms.

simultaneously at different levels. It goes without saying that these processes can be accelerated by injection of gas at the crest or water on the edge or at the base of the pool.

A few comments on the concept of pressure maintenance in fractured reservoirs are in order. Pressure maintenance may be required to maintain high productivities or to avoid the formation of deposits within the reservoir or production equipment, but it is never an objective in itself as far as recovery is concerned. It may have negative aspects for example when it stops a mechanism such as primary depletion, or be highly inefficient, for instance when sudation cannot occur. Its positive side, the replacement of matrix oil by injected fluids, involves a finely balanced decision which will be illustrated by the Iranian reservoir Haft Kel. This pool was produced for many years by depletion: a large secondary gas cap had formed, and the base was invaded by the aquifer. A few years ago the operators estimated that sudation at the base was much more important than sudation in the gas cap, because of stronger capillary forces in the water/oil system. Before implementing a waterflood project, they initiated a thorough review of the field. The evaluation of the production history was extremely complex because convection and diffusion interfered with the volumes of oil and gas. Nevertheless it was concluded that sudation in the gas cap is more effective, and this has now been explained in terms of the wettability of the matrix. The driving gravity forces in sudation are stronger in the gas oil than in the water/oil system, the capillary term being negative in both cases.

Yet another aspect has now been analysed and led to the decision to increase reservoir pressure so as to reduce the capillary effects which resist the gravity forces: the threshold pressure is given by:

$$P_d = \frac{2\,\sigma_g \cos\theta}{r} \tag{13}$$

where

σ_g = the interfacial tension between oil and gas,
θ = the wetting angle,
r = pore size.

Now σ, and therefore P_d, decrease with increasing pressure (see Fig. 17) so that sudation will be more active at high reservoir pressures. This has led to one of the largest gas injection projects in the world (Ref. 13).

Recovery factors under sudation are highly dependant on particular applications: the range of 4% at Rhourde el Baguel to 18% at Haft Kel illustrates the importance of detailed evaluations based on laboratory work.

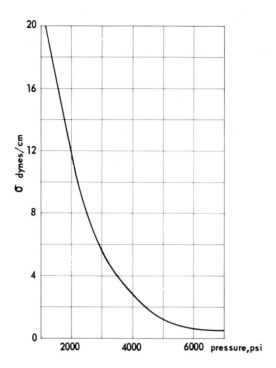

Fig. 17.   Variation of the gas/oil interfacial tension as a function
of pressure.

4.8.   CYCLIC WATER INJECTION

Cyclic water injection is designed to accelerate sudation and is carried in two phases (Refs. 18 and 19).

Phase 1: water injection

At first only the fracture network is flooded, but under continued injection part of the matrix becomes flooded as a result of increased reservoir pressure (contraction of the oil volume in the matrix, expansion of the matrix pore volume) and sudation.

Phase 2: production

During the period of pressure decline associated with production, it is hoped that oil rather than water will be preferentially expelled from water-wet rock as a result of fluid expansion and pore volume contraction.

Cyclic floods are commonly carried out below the bubble point so that a small gas saturation increases the compressibility of the matrix.

4.9. LOCALIZED DEFORMATION OF FLUID CONTACTS: CONING

The flow towards the wells is restricted to the fracture network: the cross section of pore space through which flow can take place is much smaller than in conventional reservoirs where all the cross sectional porosity is used for flow. The fluid velocity in the flow channels is about a hundred times higher than in conventional reservoirs, and turbulent flow is common near the wells as shown by the well performance which usually has a quadratic term (Appendix 2). Away from the wells, velocities are low and the high fracture permeabilities lead to very small pressure differences which are often negligeable in comparison to gravity: the fluid contacts remain parallel to their initial horizontal position (Ref. 16).

When deformation of the fluid contacts occurs, it is often due to an excessive or unbalanced rate of withdrawals. The effects such as premature breakthrough can sometimes be limited by adjusting the production rate so that gravity and sudation stabilize the movement of the fluid contacts (Ref. 20).

Coning

Coning involves the deformatin of an oil/water (gas/oil) contact due to production from a selected interval of the oil zone above the initial horizontal interface.

The description of the equilibrium between gravity and viscous forces near the well has been presented by Birks (Ref. 14) who considered the case of water and oil in an inclined fracture plane: by using Baker's (Ref. 37) laboratory and theoretical work on laminar and turbulent flow, he derived a simple formula for the pressure drawdown ΔP_{cr} at the critical rate above which water breakthrough occurs.

For a 1 000 ft drainage radius:

$$H = 1.47 \frac{\Delta P_{cr}}{\Delta \rho} \left(6.33 - 2.30 \log \frac{\Delta P_{cr}}{\Delta \rho} \right) \tag{14}$$

where

H =   distance between the base of the producing interval and the horizontal oil/water or gas oil contact, m,

$\Delta\rho$ =   gravity difference, water/oil or gas oil,

ΔP_{cr} =   pressure drawdown corresponding to the critical rate.

This relation is illustrated in Fig. 18. The critical rate is related to the draw-

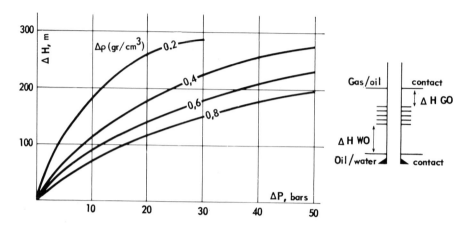

Fig. 18.   Coning in fractured reservoirs.

down by the performance equation:

$$\Delta P = AQ + BQ^2 \tag{15}$$

where A and B must be evaluated during well tests (Appendix 2) and represent semi-steady state well performance.

5

the simulation
of fractured reservoirs

The simulation of fractured reservoirs is discussed in detail in Appendix 6 which presents a wide range of mathematical models, most of which are available at Franlab's program library. These models are used to integrate the reservoir description and the analysis of production mechanisms so as to provide production forcasts.

Greater care has to be taken when analysing the results of simulation models representing fractured formations than is the case with conventional reservoirs: the input data is less reliable, the recovery processes more complex, and the equations less accurate.

We have seen that the main source of uncertainty in the basic data lies in the discontinuous nature of the fracture network, and the impossibility of measuring directly parameters which dominate reservoir behaviour. The recovery processes are more complex — this is particularly true when analysing past performance where, for instance, diffusion can undermine any attempt to keep track of gas and oil volumes. Yet another source of inaccuracy is the mathematical description of sudation.

Sudation is largely controlled by the "end effects" with which the petrophysicists who measure relative permeabilities are familiar. End effects are not yet well understood: this is discussed in Appendix 5.

It follows that the analysis of model sensitivity of the less well-known parameters is essential, and that even when some past performance is available for matching, it is worth attempting to reproduce it using alternative representations of the reservoir.

The models range from two phase, one dimensional simulators to three phases and three dimensions. Their differences lie in geometry, number of phases and recovery processes handled: in this last aspect they differ from conventional reservoir models. Most fractured reservoir simulators rely on transfer functions to describe the global behaviour, rather than calculate the interchange of fluids between matrix and fractures. These transfer functions are usually measured for the oil/water system: some models can generate their own functions which is particularly useful when laboratory work is unreliable as is the case for the gas/oil system. Hybrid models have been developed to handle specific cases such as the Emeraude field (offshore Congo) which consists of layers of low permeability rock separated by high permeability fractured beds.

The reader is referred to Refs. 15, 16, 17 which together present the "state of the art", the exception being the problem of end effects which is still a subject of active research.

6

application to the development and exploitation of fractured reservoirs

6.1. POROUS, FRACTURED RESERVOIRS

The main features of porous, fractured reservoirs whose behaviour during production has been analysed are as follows:

(a) Matrix permeability is low, of the order of one millidarcy or less.

(b) Matrix porosity is highly variable, ranging from a few per cent to 40 % in the case of some chalks.

(c) The overall permeability such as measured on well tests is highly variable — ten millidarcys upwards, values of 100 Darcys have been measured.

(d) Fracture permeability frequently falls off towards the aquifer — this is attributed to the deposition of minerals such as calcite due to the movement of subsurface waters which for obvious reasons cannot take place in the oil bearing zone after migration, and to the intensity of fracturing which can sometimes be less severe on the flanks, and in the synclines. The deposition of polar molecules of the hydrocarbons which concentrate at the oil/water contact is often found to be an impermeable barrier isolating the oil pool from its potential aquifer often referred to as a tar mat.

(e) Fracture porosity is usually a fraction of one per cent: it can be ignored when estimating oil in place.

(f) Production is usually associated with a sharp pressure decline above bubble point.

(g) Secondary gas caps usually form when reservoir pressure falls below bubble point.

(h) The injection of gas or water (sometimes in cycles) has been common practice: reservoir response has been varied.

(i) Cyclic injection is designed to accelerate sudation.

(j) Full scale water injection operations are usually implemented when sudation under capillary and gravity forces has proven to be an effective recovery mechanism: this is established by observing the efficiency of natural water influx, or by analysing the performance of pilot waterfloods. The existence of a critical rate of withdrawals can lead to much trial and error to find the optimum field offtake (Costa Foru, Ref. 20).

(k) The movement of the fluid interfaces in the fracture network can be monitored with observation wells.

(l) Anisotropic permeabilities due to preferential fracturing in one direction have been encountered in the field: this underlines the importance of early interference tests.

(m) Waterfloods of limestone reservoirs have had varied results (possibility of oil-wet matrix) (Refs. 11, 12).

(n) Table 1 illustrates the range of recoveries to be expected under different production mechanisms. They are of the order of 30%, similar to those of

TABLE 1

RECOVERY FOR TYPICAL POROUS, FRACTURED RESERVOIRS

Recovery mechanism	Field	Recovery, % oil in place			
		Achieved		Predicted	
		Average	Range	Average	Range
Dissolved gas	Kouybychev region Sprawberry Kirk, Fullerton	18	5 to 30		
Dissolved gas + Water injection	Kouybychev region Sprawberry Kirk, Fullerton	30	13 to 55		
Dissolved gas + Gas injection	Fullerton		10 to 14		
Miscellaneous : dissolved gas-gas drive with limited water influx	Mesjid i Suleiman Haft Kel	30			
Water flood above critical rate	Karabulak Achaluki Parentis			35	10 to 55

conventional reservoirs. The highest values involve depletion followed by waterflooding.,

(o) Well spacing is important, according to a statistical survey of spent Texan porous fractured reservoirs (Ref. 1):

Recovery (%)	Well spacing (m)
30	400
40	320
54	220

(p) There has been a recent revival in gas injection as witnessed by Haft Kel (Chapter 4, paragr. 4.7) and Ekofisk (North Sea).

6.2. NON POROUS, FRACTURED RESERVOIRS

The experience from the analysis of past performance of non porous, fractured reservoirs in USSR is summarized in Ref. 1:

(a) Secondary porosity, which is intimately linked with the fracture channels through which agressive waters have flowed, dissolving the matrix, is well connected.

(b) Reservoir permeability is above one darcy.

(c) Displacement efficiencies by waterflooding are extremely high, close to 100 %. This result must be treated with caution: it applies to light oils. However it has been proven by drilling wells behind the flood front, coring, and attempts to produce watered out wells, which had been shut in for long periods, at high flow rates.

(d) Depletion can lead to the formation of secondary gas caps.

(e) Recovery is unaffected by intermittent production.

(f) Secondary porosity (fractures, vugs, even caverns) are usually a few parts per thousand, sometimes as much as one per cent. This estimate is a result of cross checks of core analysis, logs, production tests and reservoir performance. These porosities probably reflect a regional character, and the following two examples will indicate the range encountered:

$$\%$$

Nagylendel . . . .   1 (Ref. 21)
Rechitsa . . . . 10 (Ref. 12)

Reliable oil in place calculations are extremely difficult to make because of the degree in uncertainty in the estimate of secondary porosity.

(g) Recovery factors listed in Table 2 are summarized below:

$$\frac{\%}{}$$

Depletion 15-20
Gas injection 60-80
Water injection 60-80

(h) Coning: common practice has been to ensure 50 m between the oil/water contact and the productive interval (Ref. 1).

(i) The oil/water contact remains close to horizontal: in the rare cases when it becomes tilted, it can be controlled by regional offtake patterns.

(j) Well spacing is very large: 1 km between production wells, 2-3 km between injection wells. Well patterns usually consist of concentric rings parallel to the structure contours.

(h) High rates of withdrawals, 10-15 % of recoverable reserves per year, are common.

These conclusions require one comment: they apply to relatively light and mobile oils (up to a few centipoises). They will not necessarily be reflected by heavy oil reservoirs such as Rospo (Italy), Nagylendel (Hungary), Emposta (Spain), which are still a subject of research and extrapolation.

TABLE 2

RECOVERY FOR TYPICAL NON POROUS, FRACTURED RESERVOIRS

Recovery Mechanism	Field	Well spacing (m)	Oil recovery, % oil in place	
			Achieved	Predicted
Pressure decline Aquifer influx, solution gas drive + Gas cap drive	Karabulak Achaluki	800	63-66	65
Pressure decline + Partial water drive	Zamankul	870		60
Depletion + Water drive	Khajan Kort	1 230	62	70

6.3. COMPLETION AND STIMULATION
OF FRACTURED RESERVOIRS

Heavy mud losses are a common feature when drilling fractured formations. In many cases the wells are not drilled to the base of the reservoir. Casing is set in the caprock, and the formation is drilled until severe mud losses indicate that a highly fractured zone which will ensure adequate productivity has been penetrated. Such wells have barefoot completions.

Improvements in drilling techniques and fluids now make it possible to drill the whole formation, and an attempt is usually made to core so as to obtain as much information as possible.

The case of barefoot as against cemented casing is not yet resolved. Barefoot completions are usually technically possible because the matrix is sufficiently compact to hold. They have the advantage of draining all the fractures, but selective completions are ruled out: the only possible workover is to plug back. When a liner or casing is run, the perforations are usually acidified to reestablish contact between the well and the fracture network. Selective completions are possible, and complex workovers can be programmed without having to run a liner.

Both types of completions have been used, and the choice will depend on local conditions. In thick fractured formations which will be waterflooded (or subject to a strong water drive), where the oil/water contact will remain close to horizontal, there may be little to be said in favour of setting a casing. On the other hand, if the same reservoir were to be depleted, casing would have to be run to protect the well from the excessive gas/oil ratios which would result from the formation of a secondary gas cap, and avoid having to run a liner across the secondary gas cap of a fractured, partially depleted reservoir.

The tendency is towards running casing, for two reasons. The first is that drilling and well completion have progressed considerably in recent years, and that casing can now be run where before it was not possible to do so. The second is a greater awareness of the need to optimize oil recovery, which means not only which fluids should be injected and how much, but also where to inject and from where to produce.

Well logging in fractured reservoirs

Logs are always run in fractured formations because the tools usually respond to the matrix properties, and have their conventional use in evaluating the lithology, porosity and water saturation of the matrix.

It is never possible to describe the fracture parameters such as matrix element size, fracture width and orientation using well logs: but well logs do react to fractures and can be used to detect their presence.

We shall review all the tools which have been used in the hope of evaluating the fracture network (see Fig. A.1.1.).

A.1.1. DIRECT METHODS

Visual logs

Both photography and closed circuit television have been adapted to well evaluation, the objective being, as it were, to enable the geologist to observe the well bore much as he would an outcrop.

Still photography has been in use for fifteen years at depths of up to 3 000 m (Ref. 12). Closed circuit television was developed later — the first major application being the Emeraude field, offshore Congo, in 1970 (Ref. 4). The television has many advantages: the camera can be focused, orientated, and moved from the surface where a standard black and white screen displays the pictures as they are being filmed: both qualitative and quantitative evaluation are possible, as shown in Fig. 10 of Chapter 2. It is possible to count the fractures, measure their orientation, vertical extension, width, etc. One important aspect is the possibility of filming the well bore during production at very low rates, when oil bleeds out of the formation: this qualitative aspect has been most useful in understanding how oil is produced from a fractured reservoir.

The technique still has severe limitations: pressure and temperature restrict its application to shallow wells (500 m) ; mud cake must be removed and the well must be filled with a clean and clear fluid.

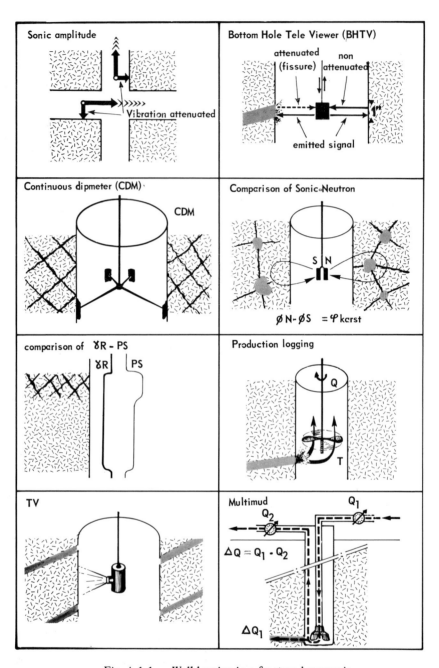

Fig. A.1.1.   Well logging in a fractured reservoir.

Bore hole televiewer (BHTV)

This tool is based on the same principle as the sonar: an ultrasonic signal scans the circumference of the well bore and the reflected energy is recorded. The response depends on well bore properties, in so far that the smoother the well bore, the stronger the reflected signal, and a vertical face is a better reflector than an inclined one. The reflected signal is treated electronically so as to give a black and white film, the lighter shades corresponding to smoother surfaces. The tool was designed in the hope of detecting irregularities (such as fractures) and changes in facies.

The tool is sensitive to well diameter and requires a well drilled to gauge. Its use is limited to muds of less than 1.15 g/cm^3 density, and it does not work in gas cut mud.

A.1.2.   METHODS INVOLVING FLOW

Record of mud losses

It is well known that mud losses often occur when drilling fractured formations: the mud cake consists of particles which are too fine to plug the fractures efficiently. Continuous recording of mud losses during drilling is becoming more widespread, and should provide an excellent cheap detector of an open fracture network. Accuracy is of the order of $15 \text{ m}^3/\text{d}$.

On exploration wells it is common practice to run tests across zones where mud losses have occured.

Production logging

Production logs, in particular flow-meter and temperature surveys, provide invaluable data. They are the best detectors of open channel networks because cemented or closed fractures make no contribution to the flow into the well bore. The intensity of fracturing of different intervals is usually reflected by their different contributions to flow:

There are two types of flow-meter survey:

Tool	Accuracy	Sensitivity to well bore diameter
Continuous flow-meter	100 m³/d	Affected
Packer flow-meter	15 m³/d	Unaffected

The continuous flow-meter is usually run in combination with a temperature survey (see Fig. A.1.2.).

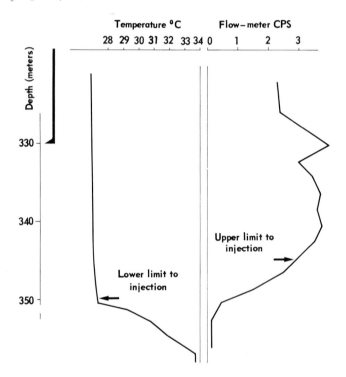

Fig. A.1.2.   Example of production logging in an injection well to determine the thickness of the fractured interval.

One common technique, used on production as well as injection wells, is to run a survey during and after an injection test. Ideally injection is carried out at several rates. The temperature survey is particularly useful. During the injection period there is little contrast on the temperature log above the strong shift which marks the lowest point of the inflow profile. When the well has been shut in, the fractured zones which have absorbed cold water show temperature anomalies as compared with the surrounding rock. When several logs are run at intervals after the well has been shut in, the rate at which each zone returns to its initial temperature can be measured and used to subdivide the formation into beds of different properties.

The advantages of the temperature survey lie in its very high sensitivity, and the fact that it is not influenced by changes in well bore diameter as is the continuous flow-meter. Its drawbacks are the lack of definition of the top of the reservoir, the impossibility of making a quantitative interpretation, and the influence of cross flow which often occurs when the well is shut in.

A marginal technique that may be mentioned under the heading "production logs" is the use of radioactive tracers which can be detected by Gamma Ray surveys: high readings indicate the intervals which have absorbed most of the injected fluid. The use of radioactive tracers presents many practical difficulties and this type of well evaluation is seldom used.

A.1.3. INDIRECT METHODS

Sonic amplitude

Theory and experiments both show that the amplitude of longitudinal and transverse waves are affected by the presence of fractures: and that maximum dampening occurs for subvertical fractures in the case of longitudinal waves, but for horizontal fractures in the case of transverse waves. This has led to attempts to measure the degree and inclination of fracturing with a sonic amplitude tool more commonly known as the cement bond log.

The sonic amplitude is very sensitive to a number of parameters which have severely curtailed its application:

(a) The reduction in signal can also result from gas cut mud or a badly centered tool: this implies that the sonic amplitude must be interpreted bearing well conditions in mind.

(b) The sonic amplitude is affected by cycle skipping particularly in carbonates.

(c) The representation of the sonic amplitude by the variable density log is often subjected to severe interference due to an increase in transit time, dampening and frequency shifts.

(d) The sonic amplitude has never been interpreted quantitatively: it is a qualitative tool.

Comparison of porosity tools

The sonic log is only affected by matrix porosity, whereas the neutron and density logs register total porosity: the difference is called "secondary porosity index" and it was once hoped that this might be a guide to fracturing. Unfortunately this has not been the case, no doubt because the accuracy of the tools, and therefore of the secondary porosity index, is too poor to detect fracture porosity values of a few parts per thousand. There is an application, however, in estimating the secondary porosities of non porous reservoirs such as karsts.

When the average karst matrix porosity is low — say 1% — it increases each time a vug or cavern of any importance is found within the perimeter being investigated by the porosity tools. For example, a vug with a radius of 10 cm corresponds to a secondary porosity of approximately 10% on Sonic and Neutron readings. It is thus possible to measure the karst secondary porosity using conventional tools (Ref. 40).

Comparison of resistivity tools

In carbonate gas bearing reservoirs with low matrix porosity, a difference has been observed between shallow (Proximity, Shallow Laterolog 9) and deep (Deep Laterolog 9, Induction Log) resistivity readings. This is a direct result of invasion and the shift can be as high as 1 log cycle (hundred to a thousand ohm-m).

Severe invasion, as calculated from logs, should always be regarded as a possible indication of fracturing.

Comparison of microresistivity logs at different angles surrounding the well bore

The best known of these tools is the continuous dipmeter (CDM).

The standard dipmeter interpretation consists of grouping the dip measurements by "shapes" which are representative of sedimentological features, formation dip, etc. There are always some apparently random values, and it was thought that they might be correlated with fracturing besides such features as geological unconformities.

The CDM results as far as the identification of fractures is concerned have so far been disappointing. Two new tools have recently been developed on similar lines:

(a) *Schlumberger's* "Fracture Identification Log" based on the correlation and superposition of the CDM curves two by two: a separation is supposed to indicate fractures.

(b) The circumferential microsonic (CMS). This new tool developed by *Schlumberger* uses four pads. The sonic wave travels around (and not along as in conventional sonic logs) the well bore. This tool should respond effectively to vertical fractures.

Combination of Gamma Ray and Spontaneous Potential (SP)

According to Smekhov (Ref. 35), the SP curve often shows an anomaly in front of fractured zones. The shale base line should be confirmed by a Gamma Ray (GR), and negative values of the SP without a change in GR may indicate fractures.

A.1.4. CONCLUSION

Logging techniques can give little more than qualitative information on the fractured network: conventional logs respond to the matrix, and production logs can help to distinguish the fractured intervals and their contribution to flow capacity of the well.

It is possible that improved interpretation methods will extend the application of logs in fractured reservoirs, but the most promising tool is the closed circuit television which is at present too restricted to be of general use.

Well performance and well tests in fractured reservoirs

The basic objective of well tests in fractured reservoirs is the same as for conventional reservoirs: productivity, well bore damage, etc. According to theory, additional data such as the size of the matrix elements can be obtained. In practice, attempts at sophisticated interpretations are often disappointing. This is partly due to the accuracy of pressure recorders, and the new generation of equipment may lead to a significant improvement in this field. Another reason is that afterflow (build-up) and irregularities in flow rate (drawdown) mask the very rapid pressure reaction due to high permeability fractures: this can only be avoided by shutting in the well down hole and involves more costly operations than are usually involved in well tests.

Turbulence occurs near the well bore even in the case of liquid flow, and the evaluation of well productivity requires tests similar to those normally practiced on gas-fields.

A.2.1. STEADY STATE FLOW BEHAVIOUR

In conventional reservoirs, the relationship between flow rate and pressure drawdown is linear, the slope giving the constant productivity index of the well.

This relationship ceases to hold in the case of the flow of gas in conventional reservoirs, and of liquids in fractured reservoirs, as a result of turbulence near the well bore. Turbulence introduces an additional pressure drop, proportional to the square of the flow rate (see Fig. A.2.1). In the case of fractured reservoirs, the lower fluid pressure near the well bore can also lead to a slight reduction in fracture width and permeability, thus further increasing the drawdown.

Maidebor (Ref. 1) has presented an analysis of semi steady state behaviour in fractured reservoirs ; he concludes that fracture width dominates turbulence, and as long as it is not affected by the drawdown, well performance is represented by the following equation:

$$\Delta P = AQ + BQ^2 \qquad \text{(A.2.1)}$$

In practice the performance curve is established by measuring the draw down at several rates and plotting the results on a graph of $\Delta P/Q$ against Q, as is sometimes done for gas wells.

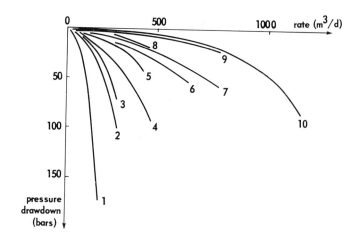

Fig. A.2.1.   Example of well performance curves.

In the case of many prolific Iranian wells, when only one point is available, an estimate of A and B is made assuming that turbulence accounts for half of the skin due to well-bore damage calculated by Pollard's method (see below):

$$BQ^2 = \frac{1}{2} \Delta P_{skin} \qquad (A.2.2)$$

The value of B can then be estimated, A being derived from Eq. A.2.1 or graphically as shown in Fig. A. 2.2.

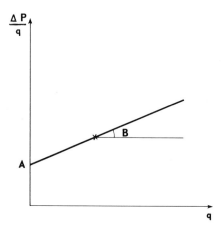

Fig. A.2.2.   Approximate calculation of well performance curve
using one point.

Baker (Ref. 37) presents analytic formulae for A and B as well as supporting laboratory measurements on a fracture of constant width separating two concrete slabs.

The first term corresponds to laminar flow, and A is the reciprocal of the productivity index:

$$A = \frac{6}{\pi} \frac{B_o \mu_o}{b^3} \ln \frac{r_e}{r_l} \tag{A.2.3}$$

The second term corresponds to the pressure losses due to viscous forces in turbulent flow and changes in kinetic energy:

$$B = \frac{\rho_o B_o^2}{4 \pi^2 b^2} \left[\frac{3}{5} \left(\frac{1}{r_l^2} - \frac{1}{r_e^2} \right) + \right.$$

$$\text{laminar flow}$$
$$\text{change in kinetic energy}$$

$$\left. \frac{\Psi}{b} \left(\frac{1}{r_w} - \frac{1}{r_l} \right) + \frac{1}{2} \left(\frac{1}{r_w^2} - \frac{1}{r_l^2} \right) \right] \tag{A.2.4}$$

$$\text{turbulent flow}$$
$$\text{viscous drag} + \text{change in kinetic energy}$$

Here:

ρ_o, μ_o and B_o are the oil density, viscosity and formation volume factor,
r is a radius,
the subscripts e, l and w refer to the external boundary, the radial limit of the region subject to turbulent flow, and the well radius,
b is the fracture width and,
Ψ is a coefficient found experimentally to be 0.011 by Baker (Ref. 37).

The changes in kinetic energy are negligible in comparison to viscous forces. When turbulence becomes important, i.e. $r_l > r_w$, the formula for B can be simplified:

$$B = \frac{1}{8\pi^2} \frac{\rho_o}{b^2} \frac{B^2}{r_w^2} \left(1 + 0.022 \frac{r_w}{b} \right) \tag{A.2.5}$$

Baker also proposes an adapted Reynolds number to be used to decide whether flow is laminar or turbulent

$$R' = \frac{B_o q_o \rho_o}{\pi r_l \mu_o} \tag{A.2.6}$$

When $R' > 6000$, turbulent flow occurs. Note that Eq. (A.2.5) can be used to estimate r_l, which is usually a few tens of centimetres, and also to determine whether turbulent flow occurs by putting $r_l = r_w$.

Injection well behaviour can differ from production well behaviour: Maidebor (Ref. 1) suggests that the injection of a cold fluid causes a contraction of the matrix elements and a widening of the fractures. This can, in some cases, more than compensate for turbulence.

A discussion of productivity decline during reservoir depletion is presented in Chapter 3: this decline is due to a reduced level of fluid pressure, and partial closing of the fractures. Productivity declines are often low because of the presence of rigid cement such as calcite within the fractures which resists the increased stresses in the reservoir.

A.2.2. TRANSIENT FLOW

Two main attempts have been made to develop methods to interpret transient flow behaviour in terms of the fracture parameters.

Pollard's method (Ref. 23)

Pollard's basic assumption, based on a simplified approach to the flow between matrix and fractures, is that the pressure drawdown is the sum of three exponential decay terms representing the fractures (rapid), well bore effects (intermediate) and matrix (slow):

$$\Delta P(t) = A_1 \exp(-\alpha_1 t) + A_2 \exp(-\alpha_2 t) + A_3 \exp(-\alpha_3 t) \quad \text{(A.2.7)}$$

$$\quad\quad\quad\quad\text{fractures} \quad\quad\quad\quad \text{well bore} \quad\quad\quad\quad \text{matrix}$$

with $\alpha_1 \gg \alpha_2 \gg \alpha_3$.

The interpretation method consists of isolating the three terms. The matrix term is obtained on a plot of $\ln \Delta P$ against time shown in Fig. A.2.3: for late times the straight line gives A_3 and α_3:

$$\ln \Delta P = \ln A_3 - \alpha_3 t \quad\quad \text{for late times} \quad\quad\quad\quad \text{(A.2.8)}$$

The third term can then be subtracted from Eq. A.2.3: for intermediate times the first term is negligible, the third has been removed, and another straight line gives $\ln A_2$ and α_2:

$$\ln \Delta P - \ln A_3 + \alpha_3 t = \ln A_2 - \alpha_2 t \quad\quad \text{for intermediate times} \quad \text{(A.2.9)}$$

According to Pollard, $\ln A_2$ corresponds to the well bore effect.

This method can be used for both drawdowns and build-ups. Unfortunately it has not had a wide range of application, and appears to be limited to particular cases: we have mentioned the evaluation of the well bore effect for Iranian

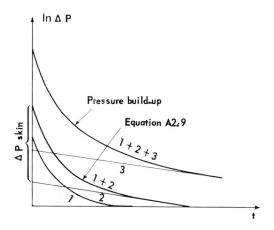

Fig. A.2.3.   Pollard's method for interpreting pressure build-ups.

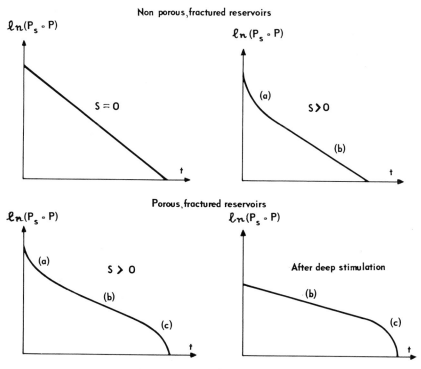

Fig. A.2.4.   Types of pressure build-up in fractured reservoirs according to Pollard.

(a) skin effect ;   (b) pressure drop within fractures ;   (c) matrix-
fracture interaction.

reservoirs (Section A.2.1 above). Chaumet *et al.* (Ref. 12) have indicated its use in distinguishing porous fractured from non porous fractured reservoirs (see Fig. A.2.4).

Method of Warren and Root (Refs. 6, 24, 25, 26)

The main theoretical drawback of Pollard's method is the simplified description of fluid flow between matrix and fractures. Warren and Root have improved upon this description: the matrix is assumed to consist of rectangular elements with orthogonal networks of fractures ; both network and matrix are assumed to be homogeneous. Semi steady state flow between fractures and matrix is assumed, and only the fractures contribute to well production.

The equations for a horizontal homogeneous isotropic reservoir can be written:

$$\frac{k_{fx}}{\mu} \frac{\delta^2 P_f}{\delta x^2} + \frac{k_{fy}}{\mu} \frac{\delta^2 P_f}{\delta y^2} - \phi_m C_{em} \frac{\delta P_m}{\delta t} = \phi_f C_{ef} \frac{\delta P_f}{\delta t} \qquad (A.2.10)$$

$$\phi_m C_{em} \frac{\delta P_m}{\delta t} = \frac{\alpha k_m}{\mu} (P_f - P_m) \qquad (A.2.11)$$

where

k_f	= fracture permeability,
P_f, P_m	= pressures in the fracture and matrix,
μ	= viscosity,
C_{ef}, C_{em}	= total effective compressibilities of the fracture and matrix,
ϕ_f, ϕ_m	= fracture and matrix porosities.

Note from Chapter 4, paragr. 4.2:

$$C_{em} = C_o + \frac{C_{pm} + C_w S_w}{1 - S_w} \qquad (A.2.12)$$

$$C_{ef} = C_o \qquad (A.2.13)$$

where

C_o, C_w and C_{pm} are the oil, water and matrix compressibilities and, S_w is the water saturation.

The first equation describes flow in the network of fractures.

The second equation represents the exchange of fluid between this network and the fractures. The parameter α is a shape factor representing the matrix geometry and properties.

A solution of these equations for an infinite reservoir producing at constant rate q is, in metric units (m³/d, m, cP, bar, d, mD):

$$\Delta P(t_D) = -\frac{q_o \mu_o B_o}{4\pi h k_f} \left[\ln t_D + 0.81 + E_i\left(-\frac{\lambda t_D}{\omega(1-\omega)}\right) - E_i\left(-\frac{\lambda t_D}{1-\omega}\right) + 2S\right]$$

(A.2.14)

where

$$t_D = \frac{k_f t}{(\phi_m \, C_{pm} + \phi_f \, C_{pf})\,\mu\, r_w^2}$$

(A.2.15)

$$\omega = \frac{\phi_f C_{pf}}{\phi_m C_{pm} + \phi_f \, C_{pf}}$$

(A.2.16)

$$\lambda = n\,(n+2)\left(\frac{r_w}{a}\right)^2 \frac{k_m}{k_m + k_f}\frac{\phi_m \, C_{pm} + \phi_f \, C_{pf}}{\phi_m \, C_{pm}}$$

(A.2.17)

where

t = time,
k_m = matrix permeability,
n = number of fracture planes: 1, 2, usually 3,
a = typical dimension of an element of matrix, e.g. height.

For small times, the E_i functions can be approximated by using logarithms as when deriving pressure drawdown equations for conventional reservoirs:

for $\dfrac{\lambda t_D}{\omega(1-\omega)} \leqslant \dfrac{1}{400}$

$$\Delta P(t_D) = -\frac{q_o \mu_o B_o}{4\pi h k_f}(\ln t_D + 0.81 + 2S - \ln \omega)$$

(A.2.18)

For large times, the E_i functions cancel:

for $\dfrac{\lambda t_D}{\omega} \geqslant 3$

$$\Delta P(t_D) = -\frac{q_o \mu_o B_o}{4\pi h k_f}(\ln t_D + 0.81 + 2S)$$

(A.2.19)

Thus the drawdown plot of pressure against log time should consist of two straight lines with identical slopes joined by an S shaped curve as shown in Fig. A.2.5. The total kh is given by the slope of these lines in the usual way. The shift on the pressure axis, ΔP, can be used to estimate ω:

$$\ln \omega = -\frac{\Delta P}{m}$$

(A.2.20)

6

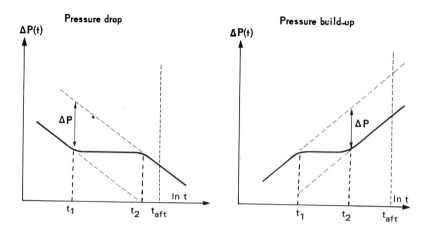

Fig. A.2.5. Transient pressure behaviour according to Warren and Root.

The compressibilities and matrix porosity being known, the value of ω can give an estimate of the fracture porosity ϕ_f.

There is considerable doubt as to validity of the S shaped portion of the curve given by Warren and Root's equation, which theoretically gives λ and therefore the block-size a. Kazemi (Refs. 34 and 6) reproduced the two straight lines using a mathematical model with a horizontal fracture network but obtains a different S shape joining them.

In practice the method developed by Warren and Root is seldom used. The pressure build-up curves obtained when testing fractured reservoirs rarely display the characteristic S shape, and the fundamental reason is that it is masked by afterflow even when the well is shut in down hole to minimize the volume of fluid active in the well. The second reason is that transient effects between fractures and matrix are not always negligible.

We shall illustrate the time scales involved by discussing formulae for the times illustrated in Figure A.2.5.

t_b = time required for semi steady state flow to prevail in the matrix,
t_1 = time at which the first straight line appears on the build-up plot,
t_2 = time at which the second straight line begins on the build-up plot,
t_{aft} = duration of afterflow.

t_1 and t_2 are given by the limits when the E_i functions can be approximated by logs, or when they cancel:

$$t_1 = \frac{1}{1\,600\,n\,(n+2)} \frac{a^2\,\mu\,\phi_f\,C_{ef}}{k_m} \tag{A.2.21}$$

$$t_2 = \frac{3}{4\,n\,(n+2)}\;\frac{a^2\,\mu\,\phi_f\,C_{ef}}{k_m}$$
(A.2.22)

t_b is discussed in Chapter 4:

$$t_b = \frac{a^2\,\mu\,\phi_m\,C_{em}}{8\,k_m}$$
(A.2.23)

t_{aft} is given by Matthews and Russell (Ref. 36):

$$t_{aft} = \frac{60\,\mu\,CV}{2\,\pi\,k_f h}$$
(A.2.24)

where

C = average compressibility of the fluid in the well contributing to after-flow,

V = volume contributing to afterflow,

h = formation thickness.

The average compressibility C lies between the gas compressibility $C_g = 1/P$ when there is gas in the well bore, and the oil compressibility C_0 if the fluid is still above bubble point.

It follows that:

$$\frac{t_1}{t_b} = \frac{1}{200\,n\,(n+2)}\;\frac{\phi_f}{\phi_m}\,\frac{C_{ef}}{C_{em}}$$
(A.2.25)

$$\frac{t_2}{t_b} = \frac{6}{n\,(n+2)}\;\frac{\phi_f}{\phi_m}\,\frac{C_{ef}}{C_{em}}$$
(A.2.26)

$$\frac{t_1}{t_{aft}} = \frac{6.55 \times 10^{-5}}{n\,(n+2)}\;\frac{C_{ef}}{C_w}\,\frac{k_f}{k_m}\,\phi_f\,\frac{ha^2}{V}$$
(A.2.27)

$$\frac{t_2}{t_{aft}} = \frac{7.85 \times 10^{-3}}{n\,(n+2)}\;\frac{C_{ef}}{C_w}\,\frac{k_f}{k_m}\,\phi_f\,\frac{ha^2}{V}$$
(A.2.28)

A typical case when afterflow effects are small – drill stem test and single-phase flow – would be:

n = 3 (3 fracture planes),

C_{em} = 0.5 x 10^{-4} bar^{-1},

C_{ef} = 10^{-3} bar^{-1}

$C = C_0$ = 1.0 x 10^{-4} bar^{-1},

k_f = 1 000 mD,

k_m = 0.1 mD,

ϕ_f $= 0.001$,
ϕ_m $= 0.1$,
V $= 1$ m^3 (40 m of 7″ below the packer),
h $= 40$ m,
a $= 30$ cm.

Substituting, after conversion into consistent units :

$$\frac{t_1}{t_b} = 10^{-4}$$

$$\frac{t_2}{t_b} = 10^{-2}$$

$$\frac{t_1}{t_{aft}} = 10^{-3}$$

$$\frac{t_2}{t_{aft}} = 10^{-1}$$

The following conclusions can be drawn:

(a) t_1 and t_2 are smaller than t_b which conflicts with the assumption that transient flow in the matrix can be neglected.

(b) t_1 and t_2 are so much smaller than t_{aft} that there is usually little hope of observing the first slope and the S shaped portion of the curves.

In practice all that would be obtained in our example is the second slope, which would be interpreted using conventional formulae to give the fracture permeability.

A.2.3. INTERFERENCE TESTS

Interference tests are particularly important in fractured reservoirs. The qualitative information is invaluable: continuity and anisotropy of the fracture network. In non porous, fractured reservoirs they can be used to estimate fracture porosity.

Porous, fractured reservoirs

We shall base our discussion on Eq. A.2.14 established by Warren and Root, modified so that the dimensionless time takes into account the spacing between the production and observation well rather than the well radius. According to Kazemi (Ref. 34), two cases occur depending on the values of λ and k_m :

(1) when $\lambda > 10^{-6}$ and $k_m > 0.1$ mD:

The fractured reservoir behaves as a conventional reservoir with:

(a) Permeability: $k_f + k_m \simeq k_f$.

(b) Porosity x compressibility : $\phi_f C_{ef} + \phi_m C_{em} \simeq \phi_m C_{em}$

(2) when $\lambda < 10^{-6}$ and $k_m < 0.1$ mD:

Interference tests reflect the size of the matrix elements and fracture porosity.

In practice many porous, fractured reservoirs can be identified with case (1) and their interference tests are analysed conventionally.

For such reservoirs, with typical values of $n = 3$ (three orthogonal planes of fracturing), $r_w = 0.1$ m, $k_m = 1$ mD:

$$\lambda = n (n + 2) \left(\frac{r_w}{a}\right)^2 \frac{k_m}{k_f} > 10^{-6}$$

if

$$k_f < \frac{150.000}{a^2} \text{ mD}$$

which is generally true even for large matrix element dimensions of the order of 1 m.

Thus interference tests in porous fractured reservoirs can often be interpreted using standard methods developed for conventional reservoirs.

Non porous, fractured reservoirs

No simple method has yet been developed to interpret interference tests in non porous, fractured reservoirs, although it is known that the results may reflect fracture porosity which is the parameter of most interest and the most difficult to ascertain even though the fluid flow through fractures and vugs is easier to describe than through a porous medium.

It is however interesting to mention the results obtained on the karstic field Nagylendel in Hungary by means of pulse testing. The porosity is calculated directly from the test using the known compressibility of the matrix for the whole of the fractured rock system. A value of about 1% was found which was completely different from the matrix porosity but close to the possible secondary porosity (Réf. 21).

A.2.4. CONCLUSIONS

Production tests are required to evaluate well productivity, which is affected by turbulence near the well bore. Transient pressure behaviour and interference tests in porous fractured reservoirs can be analysed using conventional formulae to obtain permeability and capacity. The need to demonstrate continuity of the fracture network and to discern anisotropy emphasizes the value of interference tests. Interpretation methods specific to fractured reservoirs are seldom used in practice, for operational reasons (afterflow), lack of accuracy of pressure recorders (a constraint removed by the latest generation of equipment available) and the limits of the simplified theories involved.

Nota

As this textbook was in its final printing stage, a new method has become available (Ref. 45) based on type curve analysis and taking into account afterflow (well bore storage in formation).

This method which requires accurate very early time measurements might give access to new information on fissure volume and geometry.

Relationship
between the fracture parameters

A typical illustration of a fractured reservoir is shown in Fig. A.3.1. It represents one of the many networks that can result from geological evaluation. The parameters such as matrix size and fracture width are not well

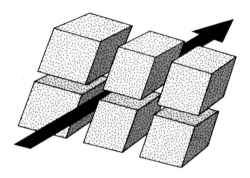

Fig. A.3.1.   Simplified fracture network.

defined by direct observation, but a good estimate of permeability can be made from well tests. Fracture permeability and porosity (k_f and ϕ_f), matrix size a and fracture width b are related, and our objective will be to derive the relationships between these parameters for different simple geometric schemes encountered in practice, shown in Fig. A.3.2.:

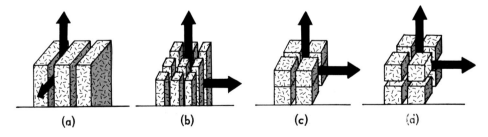

(a)   (b)   (c)   (d)

Fig. A.3.2.   Typical fracture networks — Arrows indicate possible
directions of flow.

(a) "Sheets" of matrix separated by parallel fracture planes – matrix size a is represented by the width of the sheets.

(b) "Match-sticks" separated by two orthogonal fracture planes – matrix size a is represented by the side of the square cross section.

(c) "Cubes" separated by three orthogonal fracture planes: two cases are illustrated in Fig. A.3.2. In case c the horizontal fracture is replaced by a thin stratification, a case frequently encountered in nature. In case d the three fracture systems are of equal importance. Matrix size a is represented by the side of a cubic element.

A.3.1. FRACTURE POROSITY

The rectangular element with sides a_1, a_2, a_3 is shown in Fig. A.3.3. The fracture porosity is given by:

$$\phi_f = \frac{(a_1 + b)(a_2 + b)(a_3 + b) - a_1 a_2 a_3}{(a_1 + b)(a_2 + b)(a_3 + b)} \qquad (A.3.1)$$

$$\simeq b \left(\frac{1}{a_1} + \frac{1}{a_2} + \frac{1}{a_3} \right) \qquad \text{since } b \ll a_1, a_2, a_3 \qquad (A.3.2)$$

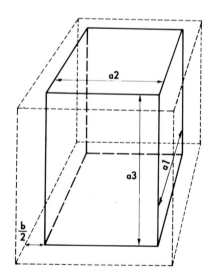

Fig. A.3.3.   Definition of fracture porosity.

For the four schemes in Fig. A.3.2.:

Sheets . . . . $\phi_f = b/a$
Match-sticks . . . . $\phi_f = 2\,b/a$
Cubes with two effective fracture planes . . . . . . $\phi_f = 2\,b/a$
Cubes . . . . $\phi_f = 3\,b/a$

A.3.2. FRACTURE PERMEABILITY

Permeability depends on the direction of flow which we shall assume to be parallel to the fracture planes as shown in Fig. A.3.4.

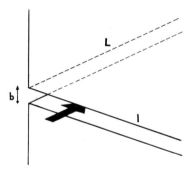

Fig. A.3.4. Definition of fracture permeability.

Poiseuille's Equation gives the rate q_1 in terms of the pressure drop ΔP for laminar flow along a single fracture whose length and cross section are L and Lb respectively:

$$q_1 = \frac{b^3 l}{12\,\mu} \frac{\Delta P}{L} \tag{A.3.3}$$

For n fractures, flow across a section A can be written:

$$q_n = n\,\frac{b^3 l}{12\,\mu} \frac{\Delta P}{L} \tag{A.3.4}$$

Darcy's law would be written:

$$q = \frac{A k_f}{\mu} \frac{\Delta P}{L} \tag{A.3.5}$$

TABLE A.3.1

Fracture Structure	Fracture network	Dimensionless				Practical Units (1)			
		f_s	ϕ_f	$k_f(\phi_f, a)$	$k_f(\phi_f, b)$	f_s (cm^{-1})	ϕ_f $(\%)$	$k_f(\phi_f, a)$ (darcy)	$k_f(\phi_f, b)$ (darcy)
Sheets		$\dfrac{1}{a}$	$\dfrac{b}{a}$	$\dfrac{1}{12}a^2\phi_f^3$	$\dfrac{1}{12}b^2\phi_f$	$\dfrac{1}{a}$	$\dfrac{1}{100}\dfrac{b}{a}$	$8,33\,a^2\phi_f^3$ (A. 3.7)	$8,33.10^{-4}\,b^2\phi_f$ (A. 3.8)
Match-sticks		$\dfrac{1}{a}$	$\dfrac{2b}{a}$	$\dfrac{1}{96}a^2\phi_f^3$	$\dfrac{1}{24}b^2\phi_f$	$\dfrac{1}{a}$	$\dfrac{1}{100}\dfrac{2b}{a}$	$1,04\,a^2\phi_f^3$ (A. 3.9)	$4,16\times10^{-4}\,b^2\phi_f$ (A. 3.10)
		$\dfrac{2}{a}$	$\dfrac{2b}{a}$	$\dfrac{1}{48}a^2\phi_f^3$	$\dfrac{1}{12}b^2\phi_f$	$\dfrac{2}{a}$	$\dfrac{1}{100}\dfrac{2b}{a}$	$2,08\,a^2\phi_f^3$ (A. 3.11)	$8,33\times10^{-4}\,b^2\phi_f$ (A. 3.8)
Cubes with one fracture plane impermeable		$\dfrac{1}{a}$	$\dfrac{2b}{a}$	$\dfrac{1}{96}a^2\phi_f^3$	$\dfrac{1}{12}b^2\phi_f$	$\dfrac{1}{a}$	$\dfrac{1}{100}\dfrac{2b}{a}$	$1,04\,a^2\phi_f^3$ (A. 3.9)	$4,16\times10^{-4}\,b^2\phi_f$ (A. 3.8)
		$\dfrac{2}{a}$	$\dfrac{2b}{a}$	$\dfrac{1}{48}a^2\phi_f^3$	$\dfrac{1}{12}b^2\phi_f$	$\dfrac{2}{a}$	$\dfrac{1}{100}\dfrac{2b}{a}$	$2,08\,a^2\phi_f^3$ (A. 3.11)	$8,33\times10^{-4}\,b^2\phi_f$ (A. 3.8)
Cubes		$\dfrac{2}{a}$	$\dfrac{3b}{a}$	$\dfrac{1}{162}a^2\phi_f^3$	$\dfrac{1}{18}b^2\phi_f$	$\dfrac{2}{a}$	$\dfrac{1}{100}\dfrac{3b}{a}$	$0,62\,a^2\phi_f^3$ (A. 3.12)	$5,55\times10^{-4}\,b^2\phi_f$ (A. 3.8)

(1) a in cm, b in microns (1 $\mu = 10^{-4}$ cm), ϕ_f in percent, k_f in darcys

so that for $q = q_n$:

$$k_f = \frac{n}{A} \frac{b^3 l}{12} = f_s \frac{b^3}{12} \qquad\qquad \text{(A.3.6)}$$

where $f_s = nl/A$ represents the total fracture length per unit cross section. For our four simplified models:

(a) Sheets $f_s = 1/a$.

(b) Match-sticks $f_s = 1/a$ or $2/a$ depending on the direction of flow: the smaller value corresponds to flow perpendicular to the axis of the matches, the larger value represents flow parallel to the axis.

(c) Cubes with two effective fracture planes: as for match-sticks.

(d) Cubes $f_s = 2/a$ for flow parallel to a fracture plane.

Note that for flow in intermediate directions, f_s would take on different values: it represents the anisotropy implied by the idealized representations of Fig. A.3.2.

A.3.3. RELATIONSHIPS BETWEEN THE FRACTURE PARAMETERS

Both porosity and permeability have been written in terms of fracture geometry, and these formulae can be rearranged so as to relate porosity directly with permeability in practical units (%, darcy) as shown in Table A.3.1. Figures A.3.5 to A.3.8. are graphical illustrations of these equations.

These graphs are of much use in estimating the possible range of values of these parameters. The permeability k_f is usually well known (from well tests), but only ranges of values of the fracture width b and element size a can be given by core analysis.

A typical application would be:

$k_f = 100$ mD	Well test
$a = 10$ cm $60 < b < 100$ (microns)	Core description

for a reservoir made up of cubic elements. Fig. A.3.8. gives a range of $30 < a < 150$ cm for the element size which contradicts the geological estimate of 10 cm. It follows that many of the fractures must be closed or sealed with cement, and that effective fracture width is lower: 40 microns, giving a fracture porosity of about 1 part per million.

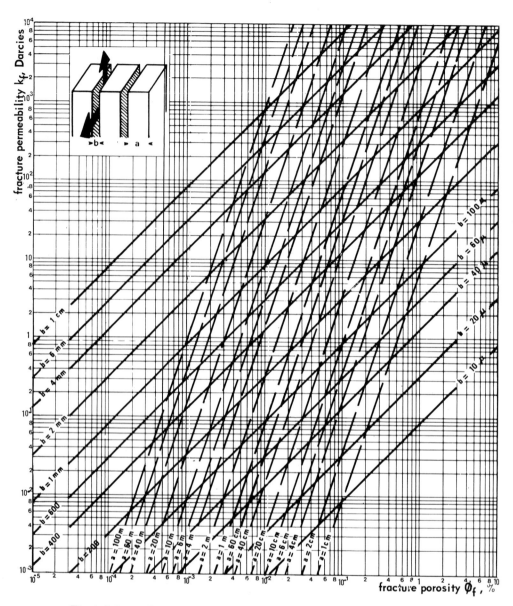

Fig. A.3.5.   Relationship between fracture permeability k_f, fracture porosity ϕ_f, fracture width b and matrix size a.

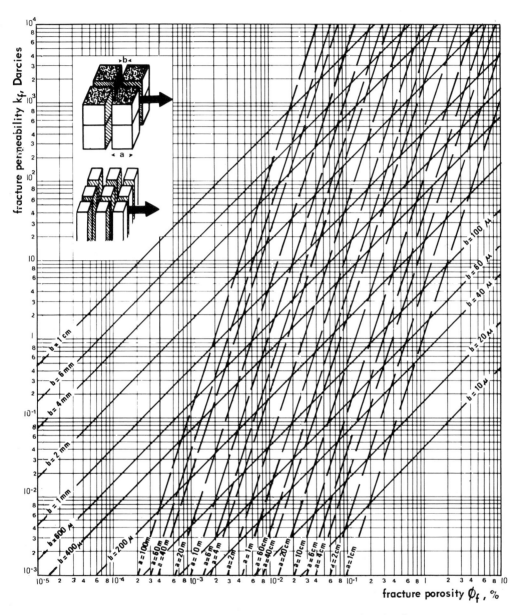

Fig. A.3.6. Relationship between fracture permeability k_f, fracture
porosity ϕ_f, fracture width b and matrix size a.

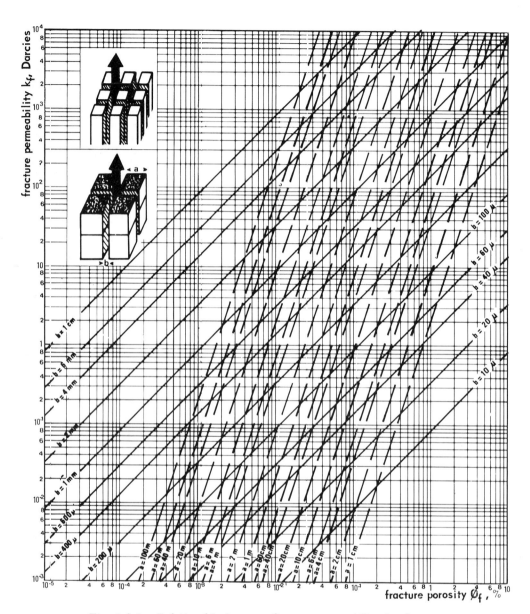

Fig. A.3.7. Relationship between fracture permeability k_f, fracture porosity ϕ_f, fracture width b and matrix size a.

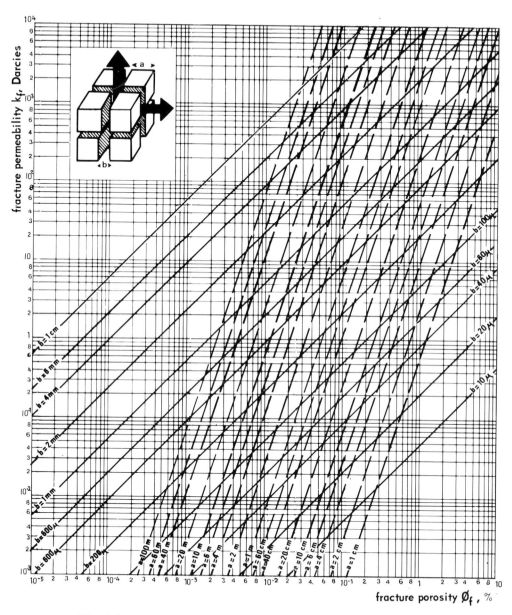

Fig. A.3.8.   Relationship between fracture permeability k_f, fracture porosity ϕ_f, fracture width b and matrix size a.

It is possible to screen parameters which lead to unusual results: thus an analysis of production performance which would suggest a 1 300 darcys reservoir with 3% fracture porosity for a reservoir whose matrix size is 10 cm would imply quite exceptional values of fracture width: 1 mm (1 000 microns).

Another method used to relate fracture porosity and permeability is the solution for radial flow, for a cubic arrangement of matrix elements whose permeability is neglected:

$$\phi_f = \frac{1}{577.9} \left\{ \frac{PI\, \mu_o\, B_o\, f_s^2 \ln \dfrac{r_e}{r_w}}{h} \right\}^{1/3} \tag{A.3.13}$$

where

ϕ_f = fracture porosity (fraction),
PI = productivity index $(m^3/d/bar)$,
μ_o = viscosity (cP),
B_o = formation volume factor (dimensionless),
r_e, r_w = drainage and well radii (cm),
f_s = parameter (cm/cm^2).

The parameter f_s has to be estimated, which implies that the size and shape of the matrix elements must be known.

A general remark is in order: these estimates of fracture porosity based on permeability are sensitive to fracture width but do not reflect secondary porosity such as vugs which may increase the effective fracture volume considerably without contributing to flow. The fracture porosities derived from Figs. A.3.6 to A.3.8 give an estimate of the minimum value compatible with the flow behaviour.

A.3.4. RELATIONSHIP BETWEEN FRACTURE PERMEABILITY AND COMPRESSIBILITY

We shall base this analysis on a system of cubic elements (side a) separated by fractures (width b).

The fracture permeability and porosity are:

$$k_f = f_s \frac{b^3}{12} \qquad \text{where} \quad f_s = \frac{1}{a} \tag{A.3.14}$$

$$\phi_f = \frac{b}{a} \tag{A.3.15}$$

The fracture compressibility is given by:

$$C_{pf} = - \frac{1}{\phi_f} \Delta \phi_f = - \frac{1}{b} \frac{\Delta b}{\Delta P} \qquad (A.3.16)$$

$$b = b_i (1 - C_{pf} \Delta P) \qquad (A.3.17)$$

Note that this fracture pore compressibility is defined with respect to the fracture volume ϕ_f.

These equations can be combined to give the following relationship between permeability and compressibility:

$$\frac{k_f}{k_{fi}} = (1 - C_{pf} \Delta P)^3 \qquad (A.3.18)$$

As mentioned in Chapter 3, productivity declines corresponding to the usual range of values of C_{pf}, i.e.:

$$10^{-3} \text{ to } 10^{-4} \text{ bar}^{-1}$$

are relatively low: 1.5 to 15% for a 50 bars pressure drop. This results from the small values of fracture compressibility and is attributed to the presence of cement maintaining the fractures open and resisting the increased effective stress on the rock.

Compressibility of fractured reservoirs

The rock compressibility that is used when dealing with conventional reservoirs reflects the deformation of the pores and not the reduction of volume of the matrix which is negligible in comparison. This matrix compressibility tends to be lower in the case of fractured reservoirs: the presence of fracturing reflects the rigidity of the rock which has broken rather than deformed elastically.

In the case of fractured reservoirs, the presence of fractures introduces an additional elasticity in the reservoir, which can be defined in two ways:

(a) In terms of the total rock volume, the fracture compressibility is defined by:

$$C_{ef} = - \frac{1}{\text{pore volume}} \frac{\Delta \text{(fracture volume)}}{\Delta \text{(pressure)}}$$

(b) In terms of the fracture volume, the fracture pore compressibility is:

$$C_{pf} = - \frac{1}{\text{fracture volume}} \frac{\Delta \text{(fracture volume)}}{\Delta \text{(pressure)}}$$

The relationship between these two definitions is obvious:

$$C_{ef} = \phi_f \, C_{pf}$$

An extensive discussion of the order of magnitude of the fracture compressibility is given in Maidebor's book (Ref. 1): C_{ef} takes on values between 10^{-5} and 10^{-6} bar^{-1}. This value is perhaps lower than might be expected, and the explanation lies in the presence of cement such as calcite which maintains the fractures open in spite of an increase in effective stress on the rock

The presence of vugs requires a further comment. We have treated vugs connected to the fracture network as part of the system of flow channels, and the definition of fracture compressibility is intended to include both the fractures and the vugs. Maidebor (Ref. 1) presents the case of a non porous, fractured reservoir for which it was possible to distinguish vug from fracture compressibility, by analysing pressure decline in the absence of water influx. In practice the vugs have to be treated as part of the matrix or as part of the fractures, and if we have chosen the latter definition, it is essentially to remain consistent with fluid flow (the vugs do form part of the flow channels) and the definition of non porous, fractured reservoirs containing oil bearing vugs.

The effective oil compressibility of the total system consisting of oil, matrix connate water, matrix and fractures, is the sum of their individual contributions, bearing in mind the relationship between the reference volume of each compressibility (e.g. the water volume for water compressibility) and the volume of oil:

$$C_{et} = C_o + C_w \frac{\phi_m S_{wm}}{\phi_m (1 - S_{wm}) + \phi_f} + C_{pm} \frac{\phi_m}{\phi_m (1 - S_{wm}) + \phi_f}$$

$$+ C_{pf} \frac{\phi_f}{\phi_m (1 - S_{wm}) + \phi_f} \qquad (A.4.1)$$

where

ϕ_m, ϕ_f = matrix and fracture porosities,
S_{wm} = matrix water saturation,
C_o = oil compressibility,
C_w = water compressibility,
C_{pm} = matrix pore compressibility,
C_{pf} = fracture pore compressibility.

In the case of porous, fractured reservoirs, $\phi_f \ll \phi_m$ and $C_{pm} \simeq C_{pf}$ so that Eq. A.4.1 becomes:

$$C_{et} = C_o + \frac{C_w S_{wm} + C_{pm}}{1 - S_{wm}} \qquad (A.4.2)$$

The fracture compressibility can be neglected in this case. A typical example:

$$S_{wm} = 0.25$$
$$C_w = 0.5 \times 10^{-4} \text{ bar}^{-1}$$
$$C_{pm} = 0.5 \times 10^{-4} \text{ bar}^{-1}$$
$$\phi_f = 0.001$$
$$\phi_m = 0.1$$
$$C_o = 1 \times 10^{-4} \text{ bar}^{-1}$$
$$C_{pf} = 1 \times 10^{-4} \text{ bar}^{-1}$$

shows that the fracture compressibility term contributes $0.001 \times 10^{-4} \text{ bar}^{-1}$ to the total effective compressibility of $1.83 \times 10^{-4} \text{ bar}^{-1}$, and confirms that it can be ignored.

In the case of non porous, fractured reservoirs, the fracture compressibility cannot be ignored: assuming a water bearing impermeable matrix ($S_{wm} = 1$), Eq. A.4.1 becomes:

$$C_{et} = C_o + \frac{C_w \phi_m + C_{pm} \phi_m + C_{pf} \phi_f}{\phi_f} \qquad (A.4.3)$$

An order of magnitude of the contribution made by the fracture compressibility in a typical case with:

$$S_{wm} = 1$$
$$\phi_f = 0.05$$
$$\phi_m = 0.05$$

and the same compressibilities as in the previous example would be 1×10^{-4} bar^{-1} as compared to $C_{et} = 3 \times 10^{-4}$ bar^{-1}.

The reason lies in the very high fracture porosity, 5%, which reflects the oil bearing vugs in the matrix.

The relationship between fracture compressibility and permeability is discussed in Appendix 3.

The main result is that the permeability variations (and consequently the productivity variations) linked to the compressibility of the fractures are relatively low. They can be calculated by the following equation:

$$\frac{k_f}{k_{fi}} = (1 - C_{pf} \Delta P)^3 \qquad (A.4.4)$$

where

$C_{pf} \simeq 1$ to 10 times 10^{-4} bar^{-1}.

Multiphase flow in fractured reservoirs

A general discussion of multiphase flow problems is presented in Chapter 4: in this appendix, the mathematical analysis will be presented in greater detail, with particular emphasis on capillary phenomena which have recently been the subject of active research.

We shall distinguish between sudation, which is the combined effect of capillary and gravity forces, and imbibition, which refers to capillarity alone.

A.5.1. SUDATION BY CAPILLARITY OR CAPILLARY IMBIBITION

Capillary imbibition is (Ref. 27) the gradual restoration of the equilibrium between two phases, the original equilibrium having been disturbed.

An example is the case of an oil saturated zone suddenly in contact with the water which has swept an adjacent layer (see Fig. A.5.1a). Fractured reservoirs represent the extreme case where one of the layers has no capillary properties, i.e. the fracture system (Fig. A.5.1b). In the general case, we will try to describe the process of capillary imbibition by the capillary properties of each layer, i.e. by their capillary pressure curves (see Fig. A.5.2).

Initially, the two layers are in capillary equilibrium with water saturations S_{w1} and S_{w2}. Their capillary pressure, depending only on the elevation over the original water table, are both equal to p_{ci}. Let us suppose that layer 1, more permeable is swept by water before layer 2.

The water saturations are no longer in equilibrium. This gives rise to a contrast in capillary pressure. Equilibrium across the boundary requires:

$$P_{o1} = P_{o2}$$
$$P_{w1} = P_{w2}$$

or

$$P_{o1} - P_{w2} = P_{o2} - P_{w2}$$

i.e.

$$P_{c1} = P_{c2}$$

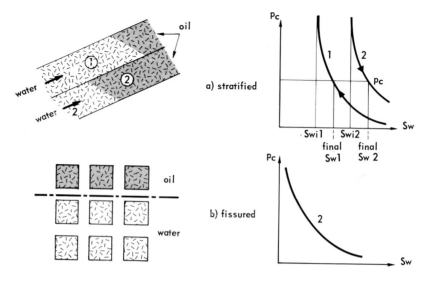

Fig. A.5.1.   Fig. A.5.2.

Capillary imbibition.

A new equilibrium will be established across the front: the swept layer 1 will lose water to the unswept layer 2, which will lose oil to layer 1. The oil thus extracted from the least permeable layer may flow towards the well through the more permeable layer. It is this we call recovery by capillary imbibition.

In the case of a fractured reservoir, layer 2 is the matrix block, represented by its capillary pressure curve (see Fig. A.5.2b), layer 1 is the fracture system, where capillary forces are negligible $(P_c = 0)$([1]). The fissures will lose water to the matrix, which will consequently lose oil until the capillary pressure becomes very low (in principle the matrix will finally contain a continuous water phase with the residual oil droplets).

In theory, the system of equations which describe the phenomenon may be established and solved numerically (Ref. 32). The results should, in any case, be used with caution, partly due to the doubt one may have as to how representative the relative permeability and capillary pressure curves might be, partly because the boundary conditions are difficult to establish. For these reasons, one still uses the direct measures in the laboratory of capillary imbibition curves (transfer function), to be transformed by dimensional analysis.

However, as we shall see, this method is also uncertain.

([1]) It should, however be noted that the imbibition is independant of the thickness of the fissure b only for $b > 25\ \mu$. See Ref. 44.

A.5.2. SUDATION: COMBINATION OF CAPILLARITY AND GRAVITY

The mathematical approach which we shall present is due to Birks (Ref. 28) and Bossie-Codreanu (Ref. 7). They represent the sudation of an element of matrix shown in Fig. A.5.3, which is partially immersed in water. We shall assume piston-like vertical displacement: water flows into the matrix at its base, and oil flows out through the top.

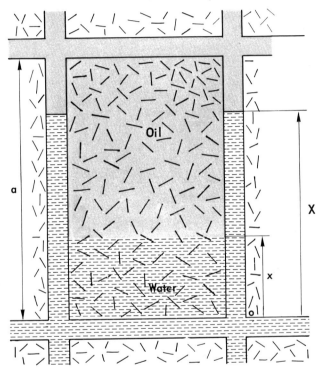

Fig. A.5.3.   Sudation as analysed by Birks.

The difference in pressure which leads to this displacement is the combined effect of capillary pressure P_c, which is constant for piston-like displacement, and the gravity term:

$$\Delta P = (X - x) (\rho_w - \rho_o) g + P_c \qquad (A.5.1)$$

where

ρ_w, ρ_o are the water and oil specific gravities.

The flow rate per unit cross section is given by Darcy's law:

$$q = \frac{k_w \, \Delta P_w}{\mu_w \, x} = \frac{k_o \, \Delta P_o}{\mu_o \, (a-x)} \tag{A.5.2}$$

where

$$\Delta P_w, \Delta P_o = \text{pressure drops in water and oil bearing zones,}$$
$$k_w, k_o, \mu_w, \mu_o = \text{permeabilities and viscosities of the two phases.}$$

It follows that:

$$\Delta P = \Delta P_w + \Delta P_o = q \left[\frac{\mu_w}{k_w} x + \frac{\mu_o}{k_o} (a - x) \right] \tag{A.5.3}$$

The velocity of the oil/water contact in the matrix is related to the flow rate by:

$$q_w = (1 - S_{cw} - S_{or}) \, \phi_m \, \frac{dx}{dt} \tag{A.5.4}$$

with S_{cw}, S_{or} the connate water and residual oil saturations in the matrix.

Combining these results leads to a differential equation for x (assuming $X = a$):

$$\frac{dx}{dt} \left[\left(\frac{\mu_w}{k_w} - \frac{\mu_o}{k_o} \right) x + \frac{\mu_o}{k_o} a \right] + \frac{g \, (\rho_w - \rho_o)}{(1 - S_{cw} - S_{or}) \, \phi_m} x = \left[\frac{(\rho_w - \rho_o) g + \frac{P_c}{a}}{(1 - S_{cw} - S_{or}) \, \phi_m} \right] a \tag{A.5.5}$$

We shall present the general solution in terms of the recovery factor:

$$E_t(\%) = 100 \, \frac{x}{a} \, \frac{(1 - S_{cw} - S_{or})}{(1 - S_{cw})} \tag{A.5.6}$$

The solution (assuming $X = a$) is given by:

$$t = - \epsilon_1 \, E_t - \epsilon_2 \, \ln \left(1 - \frac{E_t}{\epsilon_3} \right) \tag{A.5.7}$$

where

$$\epsilon_1 = \frac{a \, \phi_m}{100} \left(\frac{\mu_w}{k_w} - \frac{\mu_o}{k_o} \right) \frac{(1 - S_{wr})}{(\rho_w - \rho_o) g} \tag{A.5.8}$$

$$\epsilon_2 = a \phi_m \left(\frac{\mu_w}{k_w} - \frac{\mu_o}{k_o} \right) (1 - S_{wi} - S_{or}) \frac{\left[(\rho_w - \rho_o)g + \frac{P_c}{a} \right]}{(\rho_w - \rho_o)^2 \, g^2} \tag{A.5.9}$$

$$\epsilon_3 = 100 \, \frac{(1 - S_{wi} - S_{or})}{(1 - S_{wi})} \frac{\left[(\rho_w - \rho_o)g + \frac{P_c}{a} \right]}{(\rho_w - \rho_o) g} \tag{A.5.10}$$

This solution shows that the time required to reach a given recovery is the combination of linear and exponential terms.

When gravity is ignored, capillary imbibition leads to the well-known formula established by Bokserman and used to analyse transfer functions (Ref. 29):

$$\frac{E_t}{E_{t\,max}} = \text{Const.}\sqrt{t} \tag{A.5.11}$$

When the mobility ratio is equal to unity, i.e.:

$$\frac{k_w}{\mu_w} = \frac{k_o}{\mu_o} \tag{A.5.12}$$

the solution becomes:

$$t = -\epsilon_2 \ln\left(1 - \frac{E_t}{\epsilon_3}\right) \tag{A.5.13}$$

$$E_t = -\epsilon_3\left(1 - \exp\left(-\frac{t}{\epsilon_2}\right)\right) \tag{A.5.14}$$

which is similar to Aronovsky's formula for transfer functions (Ref. 30). Note that the rate of recovery follows an exponential decline.

A.5.3.   TRANSFER FUNCTIONS

Transfer functions are curves which represent the quantity of oil expelled from an element of matrix as a function of time: they are usually given in terms of recovery as shown in Fig. A.5.4.

They can be obtained in three ways:

(a) As the solution of the mathematical equations representing the sudation process. This approach is not altogether satisfactory because of the arbitrary definition of the boundary conditions at the fracture/matrix interface: the physical behaviour at this discontinuity is not yet fully understood.

(b) As simplified solutions which we have presented above:
. Birk's law (Eq. A.5.7, Ref. 28) which assumes vertical piston-like displacement in a partially immersed matrix.
. Boxerman's law (Eq. A.5.11, Ref. 29) which was obtained from the dimensionless analysis of a large number of experiments and later given a theoretical foundation: it implies that gravity effects can be ignored

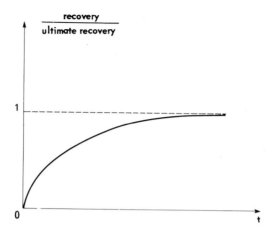

Fig. A.5.4.   Typical transfer function.

and is valid for sufficiently small matrix elements.

Aronovsky's law (Ref. 20) which assumes an exponential decline in the rate of sudation on the basis of a qualitative review of experimental data:

$$\frac{E_t}{E_{t\,max}} = 1 - \exp\left(-\frac{t}{\text{Const.}}\right) \qquad (A.5.15)$$

(c) By laboratory measurements which are scaled to field dimensions using dimensionless analysis which we shall discuss in the next section.

A.5.4.   LABORATORY EVALUATION OF TRANSFER FUNCTIONS

The standard laboratory experiment from which transfer functions (oil recovery as a function of time) are evaluated consists of placing an oil satured sample of matrix inside a water bearing container (see Fig. A.5.5a). Because of the limited size of the sample as compared to the actual block size, gravity effects are not taken into account in this type of experience. If gravity effects are considered to be important (large block size and density contrast) the container may be placed in a centrifuge (see Fig. A.5.5b). In this case, the fluids must be chosen so that the mobility ratio is the same as in the field, and the speed of the centrifuge is controlled so that the ratio of gravity to capillary forces is also identical.

Dimensionless analysis leads to two alternative transformations of the time scale:

(a) Sudation dominated by capillarity:

$$t = \sqrt{\frac{k_m}{\phi_m}} \frac{\sigma}{\mu a^2} \text{ Const.} \qquad (A.5.16)$$

(b) Sudation dominated by gravity:

$$t = \frac{k_m}{\phi_m} \frac{\Delta \rho g}{\mu a} \text{ Const.} \qquad (A.5.17)$$

where

σ = surface tension,
$\Delta \rho$ = difference in density,
g = acceleration due to gravity.

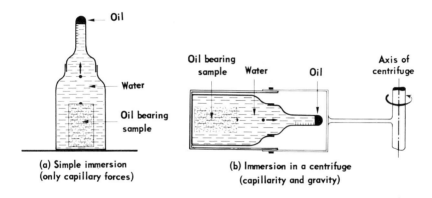

(a) Simple immersion
(only capillary forces)

(b) Immersion in a centrifuge
(capillarity and gravity)

Fig. A.5.5.   Laboratory measurement of transfer functions.

    A first drawback of these transformations is that the choice of transformation is somewhat arbitrary. A second problem is that experiments carried out with identical dimensionless quantities do not systematically lead to identical results, i.e. the scaling laws do not appear to be valid (Refs. 31, 33). One possible explanation is that the boundary conditions assumed when solving the flow equations are not correct, and this seems to be confirmed by qualitative experiments (see Fig. A.5.6, Refs. 31, 33) from which it would appear that sudation is far from uniform along the face of an element of matrix.

    The correct scaling of laboratory results is not yet resolved, and until it has been, the prediction of sudation will remain uncertain. Nethertheless there is a tendency to use numerical calculations of transfer functions (with likely boundary conditions).

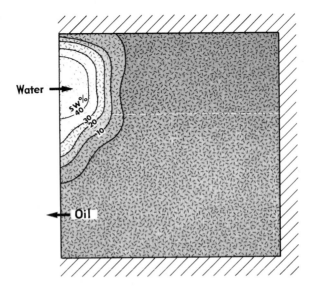

Fig. A.5.6.　Qualitative experiment of cross-flow between matrix and fracture.

Mathematical simulation
of fractured reservoirs

We shall mainly review the models in the program library belonging to the consultants *Franlab* (France) which are commercially available. They illustrate the range of tools that have been developed over the years.

From a mathematical point of view, the models can be subdivided into two categories.

The first provides a mathematical description of the interchange of fluids between matrix and fracture: a typical element is subdivided into blocks, and the flow within the element and between the edge blocks and the fractures is calculated from the equations of flow and appropriate boundary conditions. In practice, this type of model can be used to generate transfer functions. They are often used with the gas/oil system for which laboratory methods give dubious results because of the problem of scaling the gravity term correctly.

The second type of model uses transfer functions to evaluate the exchange between matrix and fractures. The reservoir is usually subdivided into blocks, and to each block is associated a set of fracture and a set of matrix properties.

An important point to bear in mind when simulating fractured reservoirs is the inaccuracy in the input data: at any given stage of reservoir development, the fracture parameters are much less well known than the properties of conventional reservoirs. When no reservoir history is available for matching, a sensitivity analysis is of the utmost importance. Even when past performance can be evaluated, the more complex recovery processes can often lead to several alternative sets of parameters. For example the gas/oil contact is measured in the fracture network, and does not indicate the total volume of gas liberated from the oil, part of which is held within the matrix: the alternatives of rapid sudation and low oil in place might both account for the past field performance but lead to widely different predictions. As a general rule it is thought to be better to evaluate sensitivity with simpler models than to indulge in a large model which, for identical costs, cannot be run with as wide a range of fundamental parameters.

A.6.1.   WATER / OIL SYSTEM

W. FRAC

This is a constant pressure model designed to evaluate waterfloods in which the oil/water contact remains horizontal, either as a result of operating conditions or of sufficiently intense fracturing. Water and oil are assumed to be segregated in the fracture network, and two sets of transfer functions are used to describe sudation depending on whether the matrix elements are partially or completely immersed in water bearing fractures.

The reservoir area is subdivided by a horizontal mesh into reservoir columns: fracture and matrix properties including matrix size are functions of depth within each column (see Fig. A.6.1).

The model is cheap, and is designed for rough preliminary evaluations and sensitivity analysis: a typical fifteen years simulation for fifteen columns uses about two minutes of CDC 6 600.

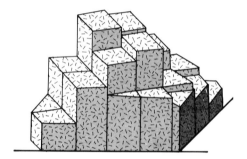

Fig. A.6.1.   Representation of a reservoir by the model W.FRAC.

FIS-2D

This is another constant pressure model designed to evaluate the deformation of the oil/water contact (see Fig. A.6.2). The geometrical representation of the reservoir is limited by a two dimensional grid. Two options are available, the radial R-Z model to evaluate coning and an X-Z version to study a cross section.

The two-phase flow in the fractures is treated supposing instantaneous segregation in the fractures. This is compatible with experimental results (see Chapter

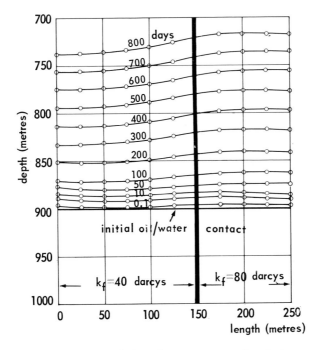

Fig. A.6.2.   Movement of the oil water contact in a heterogeneous
reservoir simulated by FIS-2D.

4.5). It leads to single-phase water flow behind and oil flow in front of the front, thus eliminating the need for relative permeability curves. Only two points are necessary: $k_{rwf} (1 - S_{orf})$ and $k_{rof} (S_{wf})$, which often, considering the low capillary forces in the fracture system, are taken equal to 1. Turbulent flow near the well bore in the radial version is not taken into account.

The matrix element size is introduced as a probability distribution, and is not constant at each point of the reservoir. Both fracture and matrix properties (including the probability distribution) vary throughout the reservoir.

Typical computing time for this model is 1.2 min. CDC 6 600 for a 800 days simulation using 200 points.

FISTUB

This model was designed for a specific application: the reservoir consists of alternating layers of highly fractured, very thin beds which act as horizontal permeability streaks or fractures, and thicker, unfractured, low permeability beds which contain most of the oil in place (see Fig. A.6.3).

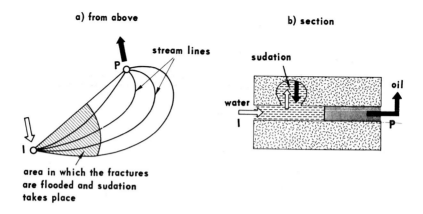

Fig. A.6.3.   Illustration of the model FISTUB.

Full pressure maintenance by waterflooding is simulated by using potential theory to calculate stream tubes in the fractured beds, and transfer functions to evaluate the sudation between the unfractured beds (matrix) and fractured beds (fracture).

An example of running costs is 30 s CDC 6 600 for one year's simulation using a 400 points model.

A.6.2. GAS / OIL SYSTEM

D. FRAC

This model is designed to evaluate the performance of a fractured reservoir during depletion, including the liberation of dissolved gas, the formation of a secondary gas cap and sudation at the top of the reservoir. The model calculates the exchange of gas and oil between matrix and fractures and does not use transfer functions.

Reservoir geometry is one dimensional and only variations of matrix and fracture properties with depth can be taken into account. This implies that the gas/oil contact is horizontal. Fractures are assumed to have infinite permeability and fracture porosity is assumed to be nil.

Figure A. 6.4. illustrates the recovery mechanisms that can be simulated simultaneously at different depths in the reservoir:

(a) Single-phase expansion where the oil is still undersaturated at the base.

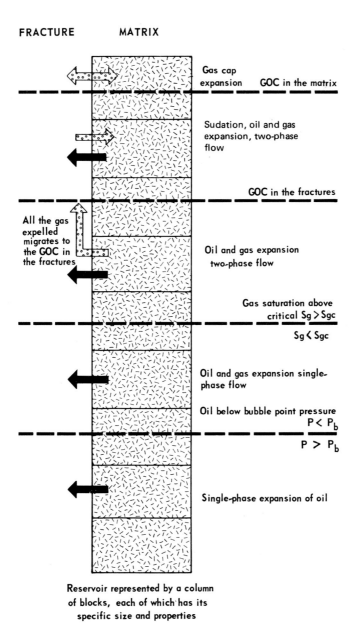

FRACTURE   **MATRIX**

Gas cap expansion   GOC in the matrix

Sudation, oil and gas expansion, two-phase flow

GOC in the fractures

All the gas expelled migrates to the GOC in the fractures

Oil and gas expansion two-phase flow

Gas saturation above critical $Sg > Sgc$

$Sg < Sgc$

Oil and gas expansion single-phase flow

Oil below bubble point pressure $P < P_b$

$P > P_b$

Single-phase expansion of oil

Reservoir represented by a column of blocks, each of which has its specific size and properties

Fig. A.6.4.   One dimensional representation of a reservoir by D. FRAC.

(b) Solution gas drive without flow of gas where the gas saturation has not reached its critical value.

(c) Solution gas drive with expulsion of gas from the matrix.

(d) Sudation and expansion of both oil and gas in the secondary gas cap.

This model can be used as a single cell to generate transfer functions for the gas/oil system.

Running costs can be relatively high owing to the time consuming calculations of the exchange of fluids between matrix and fractures: a ten year simulation of a reservoir divided into 240 layers using 5 000 calculation points costs 45 min of CDC 6 600.

G. FRAC

G. FRAC is a gas/oil version of the W. FRAC for the water/oil system (Section A.6.1). Transfer functions are normally calculated independently for input into the model, for instance by running a one cell version D. FRAC. The assumption of constant pressure when dealing with the gas/oil system restricts the use of this model to evaluating pressure maintenance by gas injection.

C. FRAC

C. FRAC is a compositional model in which the oil is represented by three components. At present, this model is restricted to a single matrix element subdivided into a large number of cells to calculate the exchange of fluids between fracture and matrix. It is designed to calculate transfer functions for a model representing reservoir geometry in greater detail.

Compositional models of fractured reservoirs which include a description of reservoir geometry are available. Core lab's YAMAMOTO is similar to D. FRAC in that the reservoir is represented by a column of grid blocks, each of which has its own set of fracture parameters: being compositional, it takes convection into account. Its cost is high for the same reason as D. FRAC – the mathematical calculation of the exchange of fluids between matrix and fractures – to which is added the additional complication of the compositional aspect.

A.6.3.   THREE-PHASE SYSTEM

FRAC TRI

FRAC TRI is a general purpose fractured reservoir model based on a three dimensional description of the reservoir geometry, and the movement of gas, oil and water.

This model has its origin in a conventional three-phase three dimensional model. The reservoir is subdivided into grid blocks which represent the fracture network: fluid flow throughout the reservoir fracture network is calculated using relative permeabilities (usually straight lines).

To each grid block (fracture network) is assigned a matrix volume with its own specific properties: matrix size, porosity, etc.; the exchange of fluids between the grid block (fracture network) and its corresponding volume of matrix elements is controlled by two sets of transfer functions, one for the water/oil system and one for the gas/oil system.

Typical recovery mechanisms that can be simulated are sudation due to water drive (the oil in the fractures is displaced by water, and once water reaches a matrix block, sudation with its associated matrix occurs) and the formation of a secondary gas cap (gas bleeds out from the matrix into its associated grid block and migrates updip through the network of grid blocks representing the fractures).

An advantage of this model is its ability to integrate production constraints such as well bore performance curves. Its disadvantage is common to most simulators which attempt to represent field geometry: the mesh used to subdivide the reservoir is coarse, and as a result the description of fluid flow suffers.

Costs are similar to those of standard three-phase models: a one year simulation for 700 grid blocks uses one minute of CDC 7 600.

A.6.4.   NON POROUS RESERVOIRS

The models discussed above are designed to evaluate the performance of porous, fractured reservoirs.

Non porous, fractured reservoirs consist of a network of secondary pores containing all the hydrocarbons in place; the matrix is compact and water bearing. The vugs may range from a few millimetres to caverns several metres across, and are hydraulically connected by the fracture network which conveyed both the agressive waters and later the oil into the reservoirs (see Fig. A.6.5).

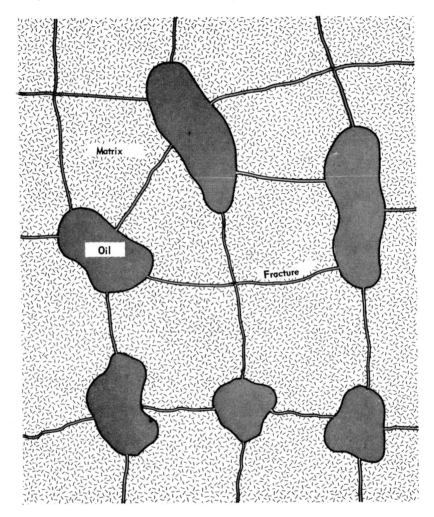

Fig. A.6.5.   Schematic representation of a portion of a karstic reservoir.

Such reservoirs are seldom encountered, and to the author's knowledge no simulation model has been written to cover this specific case. Nevertheless minor simplifications of existing simulation programs should cope with the reservoir problems associated with non porous reservoirs.

references

This list includes several papers and translations published by the *Institut Français du Pétrole* (*IFP*) and documents belonging to *Elf Aquitaine*. The author has obtained permission to provide copies of these internal documents upon request.

1 MAIDEBOR, V.N., *Production from fractured oil reservoirs.* French translation of the original Russian text, IFP 21 820 (1 and 2), Dec. 1973.

2 JEAN, F. and MASSE, P., *Methods of structural analysis.* January 1975. Original text available from the author, March 1974.

3 MASSA, D., RUHLAND, M., and THOUVENIN, J., *Structure et fracturation du champ d'Hassi-Messaoud (Algérie) (Structural geology and fracturing of the Hassi Messaoud Field in Algeria).* Published in French by Editions Technip, Paris, 1973.

4. GAY, L., and GROULT, J., *Contribution to reservoir behaviour appraisal by television. SEP AIME* preprint N° 3 749.

5. JANOT, P., *Determination of the elementary matrix block in a fissured reservoir. Application on the Eschan Field, Alsace, France. SPE AIME* preprint N° 3 638.

6 DUPUY, M., LEFEBVRE du PREY, E., and MARLE, C., *Les modèles de gisements fissurés. Etat des connaissances et des outils disponibles. (Simulation of fractured reservoirs. State of the art and models available).* In French, IFP N° 18774, Dec. 1970.

7 *Etude de quelques cas d'exploitation; Champ de Rhourde el Baguel (Algerie). (Study of fields cases. Rhourde el Baguel (Algeria).* Original French text prepared by consultants *Franlab*, July 1970.

8 SAIDI, M.A. and VAN GOLFRACHT, T., "Considérations sur les mécanismes de base dans les réservoirs fissurés" ("Discussion of basic reservoir mechanisms in fractured reservoirs"). In French, *Rev. Inst. Franç. du Pétrole,* 1971, Vol. XXVI, N° 12, p. 1167-1180.

9. COMBARNOUS, M., and BIA, P. "Combined free and forced convexion in porous media". *SPE Journal,* Dec. 1971.

10 LEFEBVRE du PREY, E., and VERRE, R., *Etude des écoulements polyphasiques dans les fissures (Study of multiphase flow in fractures).* In French, IFP N° 20950, Jan. 1973.

11 CHILINGAR, V., MANNON, W. and RIECKE, H., *Oil and gas production from carbonate rocks.* Published by Elsevier (New York, London, Amsterdam).

12 CHAUMET, P., REISS, L.H., and RUEZ, G., *Collaboration franco-soviétique sur l'exploitation des gisements. Compte rendu de mission 3-20 Oct. 1972. (Franco-Soviet cooperation on reservoir exploitation. Report on a visit 3-20 Oct. 1972).* In French, IFP N° 20939, Jan. 1973.

13 SAIDI, A.M., "Gas injection will hike recovery in Iran's gravity drainage fields". *Oil and Gas Journal,* Oct. 21st, 1974.

14 BIRKS, J., "Coning theory and its use in predicting allowable producing rates of wells in a fissured limestone reservoir". *Iranian Petroleum Institute Bulletin,* N° 12, Dec. 1963.

15 REISS, L.H., BOSSIE-CODREANU, D.N. and LEFEBVRE du PREY, E., "Flow in fissured reservoirs". *SPE AIME* preprint N° 4343.

16 SAIDI, A.M., "Mathematical simulation model describing iranian fractured reservoirs and its application to Haft Kel field". *Ninth World Petroleum Congress,* Tokyo (Japan) 1975, Pannel Discussion N° 13.

17 LEFEBVRE du PREY, E. and BOSSIE-CODREANU, D.N., "Simulation numérique de l'exploitation des réservoirs fissurés" ("Numerical simulation of the exploitation of fractured reservoirs"). *Ninth World Petroleum Congress,* Tokyo (Japan, 1975, Pannel Discussion N° 13.

18 OGANDJANJANC, V.G., *Etude au laboratoire et sur champs de l'efficacité de l'injection cyclique dans un gisement de pétrole (Laboratory and field study of the efficiency of cyclic injection in an oil pool).* French translation of the original Russian text, IFP 21443, June 1973.

19 OWENS, W.W. and ARCHER, D.L., "Waterflood pressure pulsing for fractured reservoirs". *Trans. AIME,* Vol. 237, 1966.

20 COSTA-FORU, A. and VERNESCU, A., "Comportement d'un gisement de pétrole contenu dans un réservoir calcaire fissuré de Roumanie" ("Behaviour of a Rumanian fractured limestone oil field"). In French, 1971, Vol. XXVI, N° 2, p. 81-90.

21 Hungarian document by Doleschal Sandor, available from the author.

22 DAVADANT, D., *Methodes d'étude de la fissuration par diagraphies et essais de puits (Application of logs and well tests to the study of fractured reservoirs).* In French, IFP N° 19676, Oct. 1971.

23 POLLARD, P., "Evaluation of acid treatments from pressure build-up analysis". *Trans. AIME,* Vol. 216, 1959.

24 WARREN, J.E. and ROOT, P.J., "The behaviour of naturally fractured reservoirs". *SPE Journal,* Sept. 1963.

25 BAN, A., BOGOMOLOVA, A.F., MAKSIMOV, V.A., NIKOLAIEVSKY, V.N., OGAND-JANJANC, V.G. and RYZIK, V.M., *Propriétés des roches et écoulement de filtration (Rock properties and fluid flow in porous media).* French translation of the original Russian text, IFP 11 504 and 11 505, Jan. 1965.

26 CRAWFORD, G.E., HAGEDORN, A.R. and PIERCE, E., "Analysis of pressure build-up tests in a naturally fractured reservoir". Ecole Nationale Supérieure du Pétrole et des Moteurs. *SPE AIME* preprint N° 4558.

27 MARLE, C., *Multiphase Flow in Porous Media*. Chapter V. (English Translation of the original French text. Les écoulements polyphasiques en milieu poreux. Cours de Production de l'*ENSPM*). Editions Technip, Paris, 1980.

28 BIRKS, J., "A theoretical investigation into the recovery of oil from fissured limestone formations by water drive and gas cap drive". *Fourth World Petroleum Congress.*

29 BOKSERMAN, A.A., DANILOV, V.L., JELTOV, I.P. and KOCIESKOV, A.A., *Sur la théorie de drainage des liquides non miscibles dans les roches fissurés et poreuses (Theoretical discussion of the drainage of non miscible liquids in fractured, porous media).* French translation of the original Russian text, IFP 14 925, Aug. 1967.

30 ARONOVSKY, J.S., MASSE, J. and NATANSON, S.F., "Model for mechanism of oil recovery from the porous matrix due to water invasion in fractured reservoirs". *Trans. AIME*, Vol. 213, 1958.

31 CROISSANT, R., LEFEBVRE du PREY, E., DUFORT, J. and MAULEON, M., *Effet de la gravité et de la capillarité sur la récupération de l'huile d'un bloc de réservoir fissuré (Effect of gravity and capillarity on oil recovery from an element of matrix in a fractured reservoir).* In French, IFP N° 21 566, Sept. 1973.

32 IFFLY, R. ROUSSELET, D.C. and VERMEULEN, J.L., "Fundamental study of imbibition in fissured oil fields". *SPE AIME* preprint N° 4102.

33 LEFEBVRE DU PREY, E., "Gravity and capillarity effects on imbibition in porous media". *SPE Journal*, Vol. 18, N° 3, June 1978.

34 KAZEMI, H., "Pressure transient analysis of naturally fractured reservoirs with uniform fracture distribution". *SPE Journal*, Dec. 1969, p. 463-472.

35 SMEKHOV, E.M., *Méthodes d'études des réservoirs fissurés (Methods to evaluate fractured reservoirs).* French translation of the original Russian text published by Nedra. Leningrad. Available from the author.

36 MATTHEWS, C.S. and RUSSELL, D.G., "Pressure build-up and flow tests in wells". *Monograph*, Vol. 1, Henry Doherty Series.

37 BAKER, W.J., "Flow in fissured formations" *Proceedings fourth World Petroleum Congress*, Section II-E, Paper 7.

38 RIBUOT, M., DONDON, J. and LEROY, G., *Utilisation intensive des techniques géologiques et géophysiques pour une meilleure appréciation des réservoirs (Improvement in reservoir appraisal from geology and geophysics)*, 1975. In French, available from Editions Technip, Paris.

39 ABGRALL, E. and IFFLY, R., "Etude physique des écoulements par expansion des gaz dissous" (Plysical study of dissolved gas drive). In French, *Rev. Inst. Franç. du Pétrole*, Vol. XXVIII, N° 5. p. 667-691.

40 RAPPENEAU, G., Elf Aquitaine. Personal communication.

41 Hungarian experts. Personal communication.

42 LEFEBVRE DU PREY, E., "Drainage en cascade des blocs d'un réservoir fissuré" ("Block to block interaction in fractured reservoirs"). In French, *Rev. Inst. Franç. du Pétrole*, 1976, Vol. XXXI, N° 1, p. 173-178.

43   NECTOUX, A., "Etude fondamentale de la déplétion primaire en milieu poreux avec échanges de phase". Thesis presented at *Université Claude Bernard,* (Lyon,France, 1975), Work performed at the Research Centre of *SNEA (P)* at Pau (France). The thesis concerns natural depletion in porous media with change of phase.

44   MATTAX, C.C., and KYTE, J.R., "Imbibition oil recovery from fractured water drive reservoirs". *Trans Aime,* Vol. 225, 1962.

45   GRINGARTEN, A.C., Personal communication.

Subject Index

Author Index

ACHEVE D'IMPRIMER
EN SEPTEMBRE 1980
PAR L'IMPRIMERIE LOUIS-JEAN
05002 – GAP
N° d'éditeur 499 – N° d'impression 475
Dépôt légal : 3ᵉ trimestre 1980

IMPRIME EN FRANCE